Sustainable Artificial Intelligence-Powered Applications

IEREK Interdisciplinary Series for Sustainable Development

The interdisciplinary series "Sustainable Artificial intelligence Powered Applications," (SAIPA) aims to cover a wide range of Artificial Intelligence applications with a specific focus on sustainability to further support the sustainable development goals. AI systems are able to perform a wide range of tasks, including high-level image and video recognition, perspective modeling, smart automation, advanced simulation, and complex analytics among many others, and has high learning and improvement potential through machine learning and deep learning. These capabilities have many uses and provide many benefits in all industries applications. SAIPA includes a diverse range of these applications, perspectives, and expertise, where it seeks to advance the understanding and implementation of sustainable AI-powered solutions across various scientific disciplines. Its aim is to become a comprehensive guide to different experimental sciences. The "Sustainable AI-Powered Applications" publishes conference proceedings, monographs, and textbooks that address the intersection of artificial intelligence, sustainability, and advanced technologies. Authors, researchers, contributors, and scientists from all scientific disciplines related to AI and advanced technologies, are all invited to contribute to this series to shape the future of AI in a sustainable and responsible manner. Thus, the series represents an excellent endeavor that promotes a collaborative and interdisciplinary strategy for the development of sustainable AI-driven applications. The scope of this series includes, but is not limited to: Promoting sustainability and human well-being, health and medical use, optimizing smart cities planning, transportation systems management, infrastructure development, creating livable urban environments, renewable energy, agricultural Innovations, precision agriculture, weather agriculture, sustainable food production, crop & animal monitoring, and agricultural robotics, Socio-Economic Impacts, optimizing energy consumption in buildings, factories in energy usage and lower carbon, smart grids applications, smart grids management, climate modelling, environmental monitoring, waste management, Internet of things (IOT) and other domains related to the use of AI applications that improve overall environmental impact.

Series Editor: Mourad Amer **editor@saipa-series.com**

Mohamed Gouskir · Ibtissame Aouraghe ·
Hanaa Hachimi · Belaid Bouikhalene ·
Yahya Rokni · Soufiane Belhouideg
Editors

Cutting-Edge Artificial Intelligence Applications

Springer

Editors
Mohamed Gouskir
National School of Applied Sciences
Sultan Moulay Slimane University
Beni-Mellal, Morocco

Hanaa Hachimi
École Nationale des sciences Appliquées Béni Mellal
Kenitra, Morocco

Yahya Rokni
École Nationale des sciences Appliquées Béni Mellal
Beni-Mellal, Morocco

Ibtissame Aouraghe
École Nationale des sciences Appliquées Béni Mellal
Beni-Mellal, Morocco

Belaid Bouikhalene
Faculté polydisciplinaire de Béni Mellal
Beni-Mellal, Morocco

Soufiane Belhouideg
École Nationale des sciences Appliquées Béni Mellal
Beni-Mellal, Morocco

ISSN 3005-1762 ISSN 3005-1770 (electronic)
Sustainable Artificial Intelligence-Powered Applications
ISBN 978-3-032-00124-5 ISBN 978-3-032-00125-2 (eBook)
https://doi.org/10.1007/978-3-032-00125-2

This Springer imprint is published by the registered company Springer Nature Switzerland AG
The registered company address is: Gewerbestrasse 11, 6330 Cham, Switzerland

Preface

Artificial Intelligence has evolved from being a hypothetical aspiration to a transformative force that governs the way we live, work, and interact with technology. AI has permeated nearly every domain, from revolutionizing industries to enhancing scientific discovery, offering unprecedented opportunities while simultaneously raising profound ethical and societal concerns.

AI's journey from theoretical exploration to practical implementation has been remarkable. Today, AI algorithms power everything from the recommendations we receive on streaming platforms to the autonomous vehicles navigating our cities. The sheer scale and speed at which AI is evolving necessitate ongoing debate, collaboration, and rigorous research to make certain that its potential is harnessed responsibly and effectively.

This book, a compilation of selected works from the International Conference on Cutting-Edge Artificial Intelligence Applications (CEAIA'24), highlighting the current state of AI research and its applications as it stands today. CEAIA'24, held in Beni Mellal, Morocco, on November 21–22, 2024, brought together a diverse community with a variety of backgrounds in the fields of research, industry, and practice to share the latest developments in AI and its profound impacts on various fields, from healthcare, finance, and engineering to education, and smart agriculture. The conference provided a dynamic forum for the exchange of ideas, addressing the significant issues, and envisioning the future of AI innovation.

Organized on several thematic areas, the book presents a panoramic overview of the fundamental principles, emerging trends, and applied methodologies that characterize current AI research. The book starts with the theoretical foundations of AI, setting the stage for more technical exposition in the subsequent chapters. Following this, dedicated sections explore AI's transformative role in healthcare and biomedicine, engineering and design, data science, industry, and beyond.

One distinctive feature of this collection is its interdisciplinary approach. AI is less a technical innovation but a convergence of various disciplines—computer science, mathematics, cognitive science, and ethics—all uniting to create impactful solutions in the context of a broader societal framework. The contributions to this volume highlight the potential of AI to enhance precision medicine, optimize industrial automation, redefine the face of communication and media, aid information technologies, and data science. Furthermore, as decision-making becomes more integrated with AI, transparency, fairness, and accountability issues become more urgent than ever.

Besides showcasing cutting-edge research, this book underscores the collaborative aspect of AI advancements. As the challenges' complexity that the world faces intensifies, interdisciplinary cooperation, transparent sharing of knowledge, and policy-informed AI deployment are becoming critical in order to make substantial and equitable progress through technology. The work in this volume reflects a concerted and collective effort to narrow the gap between practice and theory, providing a research landscape where AI works in the interest and for the benefit of humans.

We hope that this book provides valuable insights to researchers, students, and industry professionals who seek to better understand the evolving AI's evolving landscape. Whether through new technologies, addressing societal impacts, or developing innovative applications, the discussion in these pages contribute to the broader aim of advancing AI for the good of society.

As AI continues to push the boundaries of the possible, it becomes our collective duty and responsibility to ensure that its implementation remains ethical, inclusive, and sustainable. This book stands as both a testament of the successes that AI has already achieved and a call to action for responsible innovation in the years ahead.

Beni-Mellal, Morocco

Mohamed Gouskir
Ibtissame Aouraghe

Contents

Editors and Contributors

About the Editors

Prof. Dr. Mohamed Gouskir is an Associate Professor of Computer Science at the National School of Applied Sciences (ENSA), Sultan Moulay Slimane University in Beni Mellal, Morocco. He earned his Ph.D. in Computer Science in 2015 from the Faculty of Sciences and Techniques at the same university. His academic journey also includes a Master's degree in Business Intelligence (2011) and a Bachelor's degree in Applied Physics (2009), both obtained from Sultan Moulay Slimane University.

With a strong background in research and higher education management, Dr. Gouskir currently serves as the Coordinator of the "Artificial Intelligence and Cybersecurity" engineering program at ENSA Beni Mellal.

His research interests focus on artificial intelligence, medical image processing, and data science. He has made significant contributions to the fields of image segmentation, feature extraction, image data analysis, and classification. His expertise in using machine learning techniques for biomedical imaging has led to numerous influential publications.

Beyond academia, Dr. Gouskir is actively involved in several national and international research collaborations. He has contributed to digital transformation initiatives in education, including the deployment of online learning platforms and grade management systems at Sultan Moulay Slimane University.

As a member of various scientific committees and international conferences, he continues to drive innovation in AI applications and medical informatics. His work bridges the gap between theoretical research and practical applications, making significant strides in the field of artificial intelligence and its impact on healthcare and cybersecurity.

Prof. Dr. Ibtissame Aouraghe is a distinguished academic, researcher, and digital strategist specializing in Artificial Intelligence, and Healthcare Technologies. She is currently an Associate Professor at the National School of Applied Sciences de Béni Mellal, part of Sultan Moulay Slimane University. With a strong background in machine learning, data science, and cognitive computing, her work bridges the gap between AI research and real-world applications, particularly in the medical field.

Professor Aouraghe earned her Ph.D. in Computer Science and Intelligent Systems from Sidi Mohamed Ben Abdellah University in Fez, Morocco, in 2020. Her doctoral research focused on developing AI-driven methods for the early detection and classification of neurodegenerative diseases through online Arabic handwriting analysis. This pioneering work has been widely recognized and published in renowned international journals, demonstrating her commitment to enhancing healthcare through computational intelligence and AI-driven diagnostics.

Prior to her Ph.D., she obtained an Engineering degree in Embedded Systems and Industrial Computing from the National School of Applied Sciences at Sidi Mohamed Ben Abdellah University in Fez.

Beyond academia, Prof. Aouraghe has extensive experience in both teaching and industry. She has served as a Professor-Researcher at the Moroccan School of Engineering Sciences (EMSI—Rabat and Casablanca) and worked as a Scientific Consultant in Innovation Financing at Leyton Canada. These roles allowed her to develop strong project coordination, strategic planning, and interdisciplinary collaboration skills, contributing to the successful execution of high-impact AI and cybersecurity projects.

Prof. Dr. Hanaa Hachimi received a Ph.D. in Applied Mathematics and Computer Science from the Mohammadia Engineering School, Med V University of Rabat, Morocco, in 2013, and a Ph.D. in Mechanics and System Reliability from the National Institute of Applied Sciences of Rouen, France, in 2013. She holds a Master's degree in Mathematical Analysis and Information Processing and a Bachelor's degree in Applied Mathematics, both from the Faculty of Sciences, Ibn Tofail University of Kenitra, Morocco, in 2009 and 2007, respectively. She is a Full Professor at the National School of Applied Sciences, Ibn Tofail University of Kenitra, Morocco, and serves as President of the Moroccan Society of Engineering Sciences and Technology (MSEST).

Professor Dr. Hanaa Hachimi is Editor-in-Chief of the International Journal on Optimization and Applications (IJOA) and Director of the Laboratory of Systems Engineering (LGS). She is an IEEE Senior Member affiliated to the Big Data, Optimization, Service, and Security (BOSS) team. Professor Dr. Hanaa Hachimi is a lecturer and keynote speaker specializing in courses such as Optimization and Operational Research, Graph Theory, Statistics, Probability, Cryptography, Reliability, AI, and Scientific Computing. She is a Member of the Moroccan Society of Applied Mathematics (SM2A) and serves as General Chair of the International Conference on Optimization and Applications (ICOA) and the International Competition of Innovation (Let's Challenge). She is also a Member of the Lions Club and a UNESCO UIT Chair Member. Previously, she was the former Secretary General and Associate Professor at the Sultan Moulay Slimane University in Beni Mellal.

Prof. Dr. Belaid Bouikhalene is a distinguished Professor of Mathematics and Computer Science at Sultan Moulay Slimane University in Beni Mellal, Morocco. He obtained his Ph.D. in Mathematics in 2001 and later pursued a Master's degree in Telecommunication and Computer Science in 2007 from Ibn Tofail University, Kenitra, Morocco.

With an extensive academic and research background, Dr. Bouikhalene serves as the Director of the National School of Applied Sciences (ENSA) in Beni Mellal and leads the Digitalization Division at Sultan Moulay Slimane University. He is also the head of the Virtual Practical Work Laboratory and Director of the Laboratory for Innovation in Mathematics, Applications, and Information Theory (LIMATI).

Dr. Bouikhalene's research spans various fields, including mathematical applications, artificial intelligence, signal and image processing, information systems, business intelligence, data analysis, and e-learning. He has made significant contributions to scientific research, with over 261 publications indexed in reputable journals. He is an active reviewer for the American Mathematical Society (AMS) and MathSciNet.

Beyond academia, Dr. Bouikhalene is a recognized expert at the National Center for Scientific and Technical Research and a Member of the National Center for the Digitization of Distance Education (CNDE). He is also affiliated with the Research Group in Mathematical Inequalities and Applications (RGMIA).

Throughout his career, he has played a key role in several international research projects, including e_Omed, Tempus P@lmes, Tempus EOLES, Tempus Mission, Erasmus+ RISE, Erasmus+ MORHEL, and Erasmus+ Dialangue. His commitment to advancing digital education and artificial intelligence applications continues to shape research and innovation in these fields.

Prof. Dr. Yahya Rokni is an Associate Professor in the agro-food, currently employed at the National School of Applied Sciences of Sultan Moulay Slimane University in Béni Mellal, Morocco. Born on December 4, 1986, Yahya has developed notable expertise in the fields of bioenergetics, metabolic biochemistry, food microbiology, and industrial biotechnology.

After obtaining a Ph.D. in microbiology, biochemistry, and biotechnology from Mohammed First University in Oujda, Yahya continued his academic career as a part-time lecturer at several institutions before joining the National School of Applied Sciences in 2021. His academic work and research mainly focus on food quality and safety, as well as the improvement of biotechnological processes.

Yahya also has solid experience in quality management within the food industry, having worked for several years as a quality assurance manager in a table olive production company. He is recognized for his ability to implement quality management systems, including HACCP and ISO 22000 certifications, as well as his expertise in organic certification (ECOCERT).

Outside of his academic and professional commitments, Yahya has actively contributed to organizing scientific conferences and workshops, participating in various national and international events. He has supervised numerous final-year projects and internships for master's, bachelor's, and DUT students, thus reinforcing his role as an educator and mentor.

Prof. Dr. Soufiane Belhouideg is a Ph.D. Engineer in Engineering Mechanics and an Associate Professor at Sultan Moulay Slimane University in Beni Mellal, Morocco. His research specializes in Additive Manufacturing, smart composite structures, and the mechanical properties of materials, particularly through the lens of micromechanics.

Dr. Belhouideg has been serving as the Deputy Director of the National School of Applied Sciences Beni Mellal at Sultan Moulay Slimane University since 2023. Prior to this role, he has held various academic positions, including Associate Professor and Assistant Professor, where he taught courses in fluid mechanics, thermodynamics, wind power, and mechanical design.

He holds a Ph.D. in Mechanical and Materials Engineering from the University of Grenoble, France, and has extensive experience in both research and teaching in France and Morocco. His academic journey includes a Research Master from the University of Savoie and an Engineering degree from Mohammadia School of Engineering in Rabat, Morocco.

Dr. Belhouideg is actively involved in scientific research, having supervised multiple Ph.D. theses and led numerous projects. A robust publication record in prestigious journals and participation in significant research projects like the PHC TOUBKAL project on recycling used tires for civil engineering applications evidence his contributions to the field.

In addition to his research and teaching, Dr. Belhouideg is deeply committed to scientific outreach and innovation. He has coordinated and co-coordinated several national and international workshops and congresses, notably in 3D printing and advanced materials. He also plays an active role in academic editorial boards and as a reviewer for several international journals.

His work not only advances the understanding of material mechanics but also contributes to practical applications, such as the development of 3D-printed medical equipment during the COVID-19 pandemic.

Contributors

Mohamed Abouelmajd Research Laboratory in Physics and Engineering Sciences, Team of Applied Physics and New Technologies, Department of Physics, Polydisciplinary Faculty, University Sultan Moulay Slimane, Beni Mellal, Morocco

Abdelmoula Abouhilal Laboratory of Computer Science and Intelligent Systems, Faculty of Sciences Semlalia, Cadi Ayyad University, Marrakesh, Morocco

Houssam Abouloifa Laboratory of Bioresources, Biotechnologies, Ethnopharmacology, and Health, Faculty of Sciences, University Mohamed Premier, Oujda, Morocco;
Research Unit of Microbiology, Biomolecules and Biotechnology, Laboratory of Chemistry Physics and Biotechnology of Molecules and Materials, Faculty of Sciences and Techniques – Mohammedia, Hassan II University of Casablanca, Mohammedia, Morocco

Hamza Allaga Department of Mathematics and Informatics, Sultan Moulay Slimane University, Laboratory for Innovation in Mathematics and Applications Information Theory–LIMATI, Beni Mellal, Morocco

Mohamed Amnai Ibn Tofail University of Kenitra, Kenitra, Morocco

Ismail Arroub Research Laboratory in Physics and Engineering Sciences, Team of Applied Physics and New Technologies, Department of Physics, Polydisciplinary Faculty, University Sultan Moulay Slimane, Beni Mellal, Morocco

Abdeslam Asehraou Laboratory of Bioresources, Biotechnologies, Ethnopharmacology, and Health, Faculty of Sciences, University Mohamed Premier, Oujda, Morocco

Nadir Ayrilmis Department of Wood Mechanics and Technology, Forestry Faculty, Istanbul University-Cerrahpasa, Sariyer, Istanbul, Turkey

Ahmed Bahlaoui Research Laboratory in Physics and Engineering Sciences, Team of Applied Physics and New Technologies, Department of Physics, Polydisciplinary Faculty, University Sultan Moulay Slimane, Beni Mellal, Morocco

Mouhcine Bakhaddache Research Laboratory in Physics and Engineering Sciences, Team of Applied Physics and New Technologies, Department of Physics, Polydisciplinary Faculty, University Sultan Moulay Slimane, Beni Mellal, Morocco

Mohamed Baslam TIAD Laboratory, FST, Sultan Moulay Slimane University, Beni Mellal, Morocco

Soufiane Belhouideg Research Laboratory in Physics and Sciences for Engineers (LRPSI), Polydisciplinary Faculty, Sultan Moulay Slimane University, Béni-Mellal, Morocco;
Research Laboratory in Physics and Engineering Sciences, Team of Applied Physics and New Technologies, Department of Physics, Polydisciplinary Faculty, University Sultan Moulay Slimane, Beni Mellal, Morocco

Reda Bellaouchi Laboratory of Bioresources, Biotechnologies, Ethnopharmacology, and Health, Faculty of Sciences, University Mohamed Premier, Oujda, Morocco

Jamaa Bengourram Laboratory of Industrial Engineering and Surface Engineering, Sciences and Technology Faculty, Sultan Moulay Slimane University, Beni Mellal, Morocco;
Industrial Engineering Laboratory, Mechanical Engineering Department, Faculty of Science and Technology, University Sultan Moulay Slimane, Beni Mellal, Morocco

Belaid Bouikhalene Laboratory of Mathematics Innovation and Information Technology (LIMATI), Polydisciplinary Faculty Beni Mellal, Sultan Moulay Slimane University, Beni Mellal, Morocco

Abdelati Bourzik LIMATI Laboratory, Polydisciplinary Faculty, Sultan Moulay Slimane University, Beni Mellal, Morocco

Mohamed El Alaoui National School of Applied Sciences, Sultan Moulay Sliman University - USMS, Beni-Mellal, Morocco;
Laboratory of Physics and Engineering Sciences, Research Team in Embedded Systems Engineering, Automation, Signal, Telecommunications and Smart Materials, Polydisciplinary Faculty, Sultan Moulay Sliman University - USMS, Beni-Mellal, Morocco

Radouane El Amri Environmental, Ecological and Agro-Industrial Engineering Laboratory, Department of Chemistry and Environment, Faculty of Sciences and Technology, University Sultan Moulay Slimane, Beni Mellal, Morocco

Nour El Houda Benkaddour Laboratory of Epidemiology, Clinical Research and Public Health, Faculty of Medicine and Pharmacy of Oujda, Mohammed First University, Oujda, Morocco

Jaouad El-Mekkaoui LSEM Laboratory, Higher School of Technology, Sidi Mohamed Ben Abdellah University, Fez, Morocco

Ismail Ezzaraa Industrial Engineering Laboratory, Mechanical Engineering Department, Faculty of Science and Technology, University Sultan Moulay Slimane, Beni Mellal, Morocco

Abderrazak Farchane Department of Mathematics and Informatics, Sultan Moulay Slimane University, Laboratory for Innovation in Mathematics and Applications Information Theory–LIMATI, Beni Mellal, Morocco

Sara Gaamouche Laboratory of Bioresources, Biotechnologies, Ethnopharmacology, and Health, Faculty of Sciences, University Mohamed Premier, Oujda, Morocco

Nabil Ghabbour Laboratory of Bioresources, Biotechnologies, Ethnopharmacology, and Health, Faculty of Sciences, University Mohamed Premier, Oujda, Morocco;
Laboratory of Natural Resources and Environment, Polydisciplinary Faculty of Taza, Sidi Mohamed Ben Abdellah University, Taza, Morocco

Lahcen Gouskir TIAD Laboratory, FST, Sultan Moulay Slimane University, Beni Mellal, Morocco

Mohamed Gouskir National School of Applied Sciences, Sultan Moulay Slimane University, Béni Mellal, Morocco

Hanaa Hachimi National School of Applied Sciences, Kenitra, Morocco

Intissar Haddiya Laboratory of Epidemiology, Clinical Research and Public Health, Faculty of Medicine and Pharmacy of Oujda, Mohammed First University, Oujda, Morocco;
Department of Nephrology, Mohammed VI University Hospital, Faculty of Medicine and Pharmacy of Oujda, Mohammed First University, Oujda, Morocco

Said Hakimi Department of Mathematics and Informatics, Sultan Moulay Slimane University, Laboratory for Innovation in Mathematics and Applications Information Theory–LIMATI, Beni Mellal, Morocco

Ismail Hasnaoui Laboratory of Bioresources, Biotechnologies, Ethnopharmacology, and Health, Faculty of Sciences, University Mohamed Premier, Oujda, Morocco

Aziz Hirri National School of Applied Sciences, Sultan Moulay Slimane University, Beni Mellal, Morocco

Amal Hjouji STIC Laboratory, Faculty of Science, Sidi Mohamed Ben Abdellah University, Fez, Morocco

Nabil Hmina Laboratory of Advanced Systems Engineering, National School of Applied Sciences ENSA, Ibn Tofail University Campus, Kenitra, Morocco

Mohamed Karouchi Laboratory of Engineering in Chemistry and Physics of Matter Faculty of Sciences and Technics, Sultan Moulay Slimane University, Beni Mellal, Morocco

Zakaria Khaouch Industrial and Surface Engineering, Faculty of Science and Technology, Sultan Moulay Slimane University, Beni Mellal, Morocco

Nourreeddine Kouider Industrial and Surface Engineering, Faculty of Science and Technology, Sultan Moulay Slimane University, Beni Mellal, Morocco

Manuel Lagache Polytech Annecy Chambery, University of Savoie, SYMME, Annecy, France

Cherki Lahlou Laboratory of Industrial Engineering and Surface Engineering, Sciences and Technology Faculty, Sultan Moulay Slimane University, Beni Mellal, Morocco

Hassan Latrache Laboratory of Industrial Engineering and Surface Engineering, Sciences and Technology Faculty, Sultan Moulay Slimane University, Beni Mellal, Morocco

Saad Mahmoudi Ibn Tofail University of Kenitra, Kenitra, Morocco

Kaoutar Mchiouer Higher Institutes of Nursing and Health Technical Professions Nador (ISPITS), Nador, Morocco

Asmae Najm-Eddine Research Laboratory in Physics and Engineering Sciences, Team of Applied Physics and New Technologies, Department of Physics, Polydisciplinary Faculty, University Sultan Moulay Slimane, Beni Mellal, Morocco

Youssef Najm-Eddine Research Laboratory in Physics and Engineering Sciences, Team of Applied Physics and New Technologies, Department of Physics, Polydisciplinary Faculty, University Sultan Moulay Slimane, Beni Mellal, Morocco

Sara Ramdani Laboratory of Epidemiology, Clinical Research and Public Health, Faculty of Medicine and Pharmacy of Oujda, Mohammed First University, Oujda, Morocco

Yahya Rokni Research Unit Bioprocess and Biointerfaces, Laboratory of Industrial Engineering and Surface Engineering, FST, Sultan Moulay Slimane University, Beni Mellal, Morocco;
Laboratory of Bioresources, Biotechnologies, Ethnopharmacology, and Health, Faculty of Sciences, University Mohamed Premier, Oujda, Morocco

Meryem Samine Industrial and Surface Engineering, Faculty of Science and Technology, Sultan Moulay Slimane University, Beni Mellal, Morocco

Sara Samine Laboratory of Advanced Systems Engineering, National School of Applied Sciences ENSA, Ibn Tofail University Campus, Kenitra, Morocco

Chorouk Zanane Research Unit Bioprocess and Biointerfaces, Laboratory of Industrial Engineering and Surface Engineering, FST, Sultan Moulay Slimane University, Beni Mellal, Morocco;
Higher Institutes of Nursing and Health Technical Professions Béni Mellal (ISPITS), Beni Mellal, Morocco

Maria Zemzami National Graduate School of Arts and Crafts, Mohamed V University, Rabat, Morocco

Tracing the Evolution and Impact of Artificial Intelligence: Insights into Algorithms, Networks, and Applications

Sara Samine, Maria Zemzami, Nabil Hmina, Manuel Lagache, and Soufiane Belhouideg

Abstract

The foundation of artificial intelligence (AI) is algorithmic models and mathematical ideas that facilitate learning and automated decision-making. With an emphasis on supervised, unsupervised, and reinforcement learning paradigms, this work examines the theoretical underpinnings of machine learning and deep learning. The article explores sophisticated deep learning architectures, such as convolutional neural networks (CNNs), for computer vision and recurrent neural networks (RNNs) for sequential data processing. It also discusses autoencoders, which are employed for dimension reduction and anomaly identification, as well as generative adversarial networks (GANs) for the production of synthetic data. To conclude, the article highlights the use of AI in several sectors such as computer vision, natural language processing, healthcare, finance, and additive manufacturing, demonstrating its influence on improving processes and enhancing the performance of intelligent systems.

Keywords

Artificial intelligence (AI) · Algorithmic decision-making · Convolutional autoencoders · CNN · Computer vision

1 Introduction

Artificial intelligence is a branch of computer science whose aim is to create systems that integrate a large amount of knowledge and processing, known as intelligent systems.

It also aims to design systems capable of performing tasks traditionally requiring human intelligence, such as visual recognition, understanding natural language, decision-making, and learning.

Since its inception, AI has evolved thanks to advances in mathematics, computer science, and neuroscience, enabling the creation of sophisticated algorithms and models. The notion of AI was first mentioned in the 1950s by mathematician BY (Turing 1950; Samine et al. 2023).

Lately, artificial intelligence has become more widely used in fields such as intelligent robots, gaming, security, cybersecurity, industry, and additive manufacturing (LeCun et al. 1989, 2015a; Lopez-Jimenez et al. 2020; Agrawal et al. 2019; He et al. 2022). Thanks to advances in mathematics, computer science, and the neurosciences, it has also become more and more popular.

Deep learning is a subfield of machine learning, which is a subfield of artificial intelligence (AI).

Artificial intelligence is a discipline that encompasses several subfields. The figure below illustrates the relationship between three key concepts: artificial intelligence, machine learning, and deep learning (Fig. 1).

S. Samine (✉) · N. Hmina
Laboratory of Advanced Systems Engineering, National School of Applied Sciences ENSA, Ibn Tofail University Campus, Kenitra, Morocco
e-mail: sara.samine@uit.ac.ma

N. Hmina
e-mail: hmina@uit.ac.ma

M. Zemzami
National Graduate School of Arts and Crafts, Mohamed V University, Rabat, Morocco
e-mail: m.zemzami@um5r.ac.ma

M. Lagache
Polytech Annecy Chambery, University of Savoie, SYMME, Annecy, France
e-mail: manuel.lagache@univ-savoie.fr

S. Belhouideg
Research Laboratory in Physics and Sciences for Engineers (LRPSI), Polydisciplinary Faculty, Sultan Moulay Slimane University, Béni-Mellal, Morocco
e-mail: soufianebelhouideg@usms.ma

M. Gouskir et al. (eds.), *Cutting-Edge Artificial Intelligence Applications*, Sustainable Artificial Intelligence-Powered Applications,
https://doi.org/10.1007/978-3-032-00125-2_1

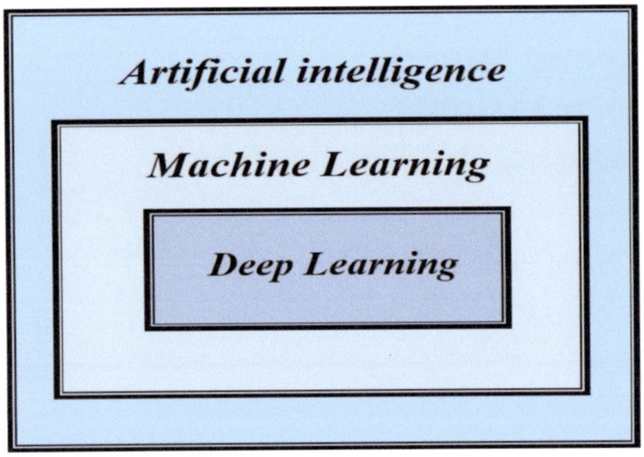

Fig. 1 The artificial intelligence landscape

Artificial intelligence generally covers all methods enabling machines to reproduce human intelligence. Machine learning is a branch of artificial intelligence that uses algorithms to train on data (Samine et al. 2022). Finally, deep learning, a branch of machine learning, uses deep neural networks to handle complex data and perform complex tasks such as image recognition and natural language processing (Li et al. 2022; Ma et al. 2022).

2 Machine Learning Techniques and Applications

Machine learning is generally divided into three categories (Wei et al. 2019; Rachna Vaish et al. 2021):

2.1 Supervised Learning

Supervised learning deals with labeled data, which means that you give the algorithm training examples where you have both input features and the target output. The model minimizes the difference between its predictions and the labels it was trained on.

Examples of Algorithms:

- k-nearest neighbors (k-NN).
- Linear and logistic regression.
- Decision trees and random forests.
- Neural networks (e.g., CNNs for image classification).

Applications:

- Medical diagnosis (predicting diseases from patient data).
- Fraud detection in financial transactions.
- Spam email classification.

2.2 Semi-supervised Learning

Semi-supervised learning is a hybrid approach that combines a small set of labeled data with a large amount of unlabeled data. This technique leverages the structure of the unlabeled data to improve model accuracy.

Examples of algorithms

- Self-training models.
- Graph-based methods (e.g., label propagation).
- Semi-supervised deep learning (e.g., semi-supervised GANs).

Applications

- Text classification when only a few labeled documents are available.
- Image recognition in medical imaging where annotation is costly.
- Speech recognition with a small labeled dataset.

2.3 Unsupervised Learning

Unsupervised learning works with unlabeled data, aiming to uncover patterns, structures, or relationships within the dataset (Samine et al. 2024).

Some examples of unsupervised learning methods are K-means and other clustering algorithms. Dimensionality reduction methods, such as PCA, kernel-PCA, and LLE, can also be used to transform the data while preserving key variance (Wang 2012). This also includes autoencoders, which learn a compressed representation of the data.

2.4 Reinforcement Learning

Reinforcement learning (RL) is a learning paradigm in which an agent is placed in an environment to learn how to achieve its goals through a system of punishment and reward (Sahu et al. 1956; Xu et al. 2022).

Key concepts

- State: The current situation of the agent.
- Action: Possible choices the agent can make.
- Reward function: Provides feedback (positive or negative) based on the action taken.
- Policy: The strategy the agent learns over time.

Examples of algorithms

- Q-learning: An algorithm based on a table of action values.
- Deep Q networks (DQN): An extension of Q-learning using neural networks.
- Policy gradient and proximal policy optimization (PPO): Policy optimization methods.

Applications

- Reinforcement learning is used in a number of advanced fields:
- Robotics, to enable robots to learn complex tasks such as manipulating objects.
- Video games and artificial intelligence: AlphaGo and agents capable of beating human players.
- Autonomous vehicles, to optimize decision-making in real time.

The various types of machine learning are shown in Fig. 2.

Machine learning is applied in several fields (Kappes et al. 2018; Liu et al. 2017):

- Computer vision: facial recognition, object detection, and image.
- Analysis.
- Natural language processing: Machine translation, virtual personal assistants, and sentiment analysis.
- Healthcare: AI-supported diagnostics, disease anticipation, and drug discovery.
- Finance: Fraud identification, risk assessment, and algorithmic trading.
- Manufacturing: Prediction and improvement of production procedures.
- Additive manufacturing: Error detection, prediction of mechanical properties of 3D printed parts.

3 Fundamental Steps in Machine Learning

To create a machine learning model, follow the steps given below (Fig. 3) (Yang et al. 2022):

- Compile and get your dataset ready.
- Divide your dataset into test and train sets.
- Adjust the hyperparameters.
- Develop your model.
- Analyze the model's performance.

4 Deep Learning

Deep learning is a sub-domain of machine learning that relies on multi-layer artificial neural networks to analyze and learn complex patterns from large amounts of data (LeCun et al. 2015b; Schmidhuber 2015).

There are several types of deep neural networks, each designed to meet specific needs:

- Convolutional Neural Networks (CNN): Mainly used in computer vision for tasks such as image recognition, object detection, and medical examination.
- Recurrent Neural Networks (RNN): Specially designed for sequential data, such as natural language processing, machine translation, and speech recognition.
- Autoencoders: Used for anomaly detection and size reduction, data compression.
- GANs (Generative Adversarial Networks): Used for image generation and quality enhancement (Mak et al. 2023).

4.1 Convolutional Neural Networks CNN

Convolutional neural networks (CNNs) are models specifically inspired by the workings of the human brain, designed to process images and automatically extract important features. Thanks to their multi-layered architecture, they can analyze images in depth and identify relevant patterns. These networks are commonly used for applications such as image classification object recognition and segmentation of different parts of an image, as illustrated in Fig. 4 (Kattenborn et al. 2021; Ketkar and Moolayil 2021).

For a convolutional neural network, there are four types of layers (Hossain and Sajib 2019):

- Convolution layer (CONV): Convolution is a common practice in image processing. It involves dragging a filter (or kernel) over an image, followed by its convolution with the latter. The convolution layer is the fundamental element of a convolutional neural network (CNN).
- Pooling layer: The aggregation layer is a variant of input image sub-sampling. It facilitates the reduction of an image while preserving its essential elements, based on the extraction of important pixel values. Max pooling is the most frequently used method.
- ReLU correction layer: The ReLU correction layer replaces all negative input values with zeros. It acts as an activation function.
- Fully connected layer: The fully connected layer is therefore the last layer of a convolutional neural network.

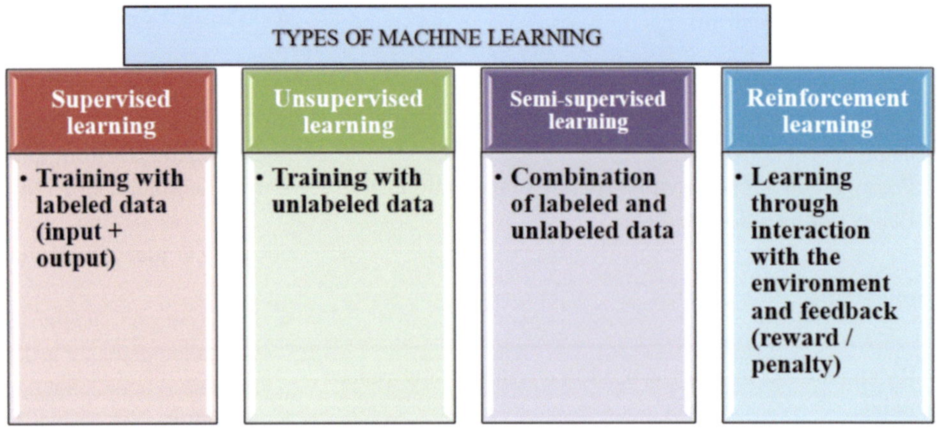

Fig. 2 Categories of machine learning

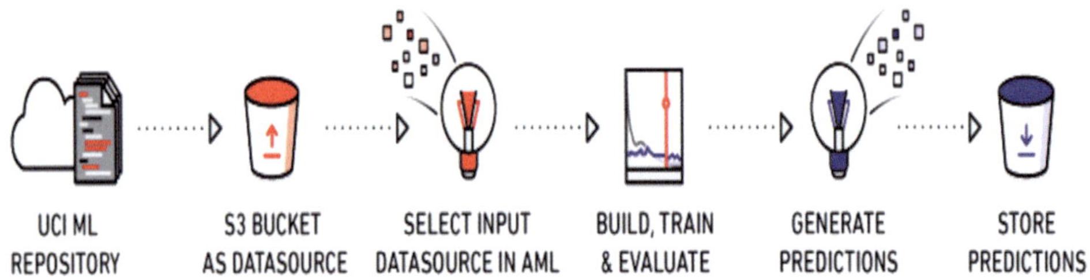

Fig. 3 Essential steps in the machine learning (image source: Amazon Web Services)

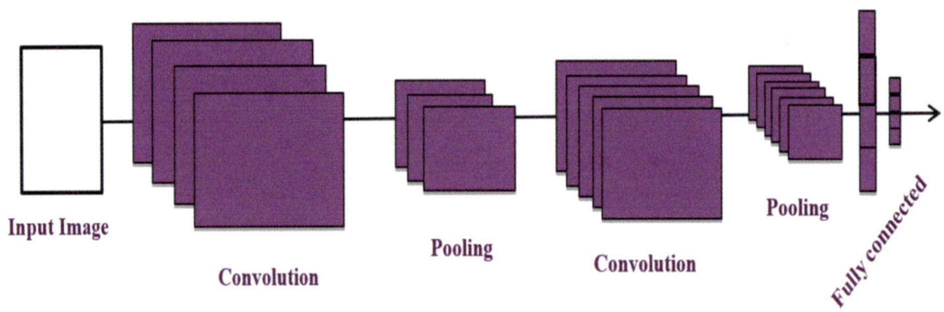

Fig. 4 Principle of a convolutional network (CNN)

4.2 Deep Learning Models Based on Recurrent Neural Networks (RNNs)

Deep learning models based on recurrent neural networks (RNNs) are a class of models designed to process sequential data. Unlike standard neural networks, which process each input independently, RNNs feature recurrent links, giving them the ability to retain data from previous phases and use it to examine subsequent phases (Yin et al. 2017; Kaur and Mohta 2019).

The figure above shows the difference between a classical neural network and a recurrent neural network.

In a classical network, information can only flow in one direction, whereas in a recurrent neural network, information can flow in both layers (Fig. 5).

4.3 Autoencoder

An autoencoder is a type of unsupervised neural network that learns to encode data into a compact representation before reconstructing it. It consists of two main components: an encoder, which transforms the input data into a lower dimensional representation, and a decoder, which attempts to reconstruct the original data from this representation (Fig. 6). Unlike clustering methods, autoencoders do not directly organize data into groups; however, they can be used for feature learning, which may facilitate clustering tasks.

Autoencoders are widely used for dimensionality reduction, anomaly detection, and image denoising. In anomaly detection, such as in cybersecurity or predictive maintenance, an autoencoder is trained on normal data and struggles to reconstruct anomalous samples, allowing for their

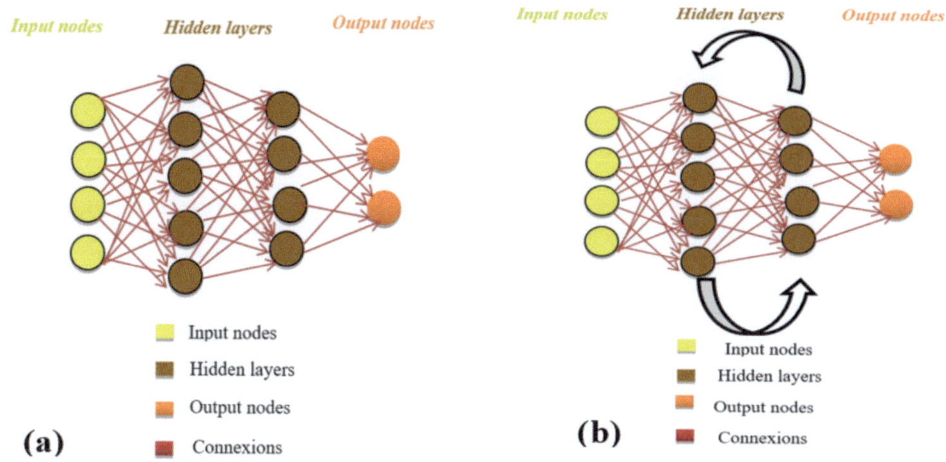

Fig. 5 **a** Classical neural network, **b** recurrent neural networks

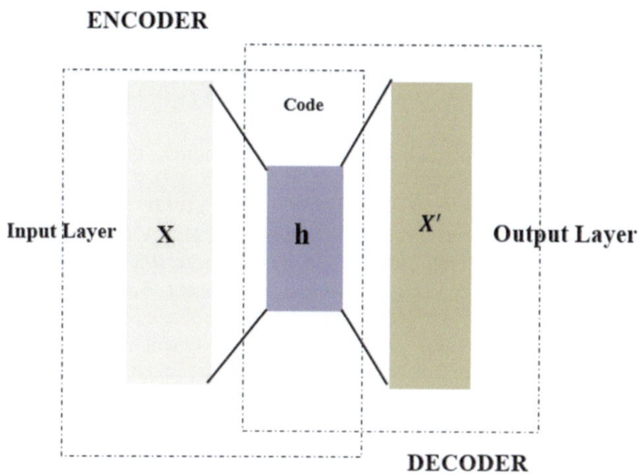

Fig. 6 Diagram of an autoencoder

identification. In imaging, they are employed to remove noise and restore degraded images by learning to reconstruct a clean version of the input. Autoencoders are also used for data compression, learning compact representations that can be more efficient than traditional compression methods (Sakurada and Yairi 2014; Wei et al. 2019).

A more advanced variant, known as a variational autoencoder (VAE), introduces a probabilistic approach to data representation. Unlike standard autoencoders, which learn fixed representations, VAEs model the underlying distribution of data using continuous latent variables, allowing them to generate new samples similar to the training data. This makes VAEs particularly useful for image synthesis, text generation, and exploring variations within complex datasets (Kaur and Mohta 2019).

5 Conclusions

Artificial intelligence has profoundly transformed many fields thanks to the rise of machine learning and deep learning. Supervised, unsupervised, and reinforcement learning enable machines to process complex data and automate tasks that were once reserved for humans. The evolution of AI, from early algorithms to advanced neural networks, has made systems more autonomous and efficient, paving the way for various applications in computer vision, natural language processing, and industrial process optimization.

Its impact now extends across numerous sectors, enhancing efficiency and precision. However, despite these advancements, challenges remain, particularly in ethics, algorithm transparency, and model explainability. AI continues to evolve rapidly, offering new opportunities and innovations that will shape the future of intelligent technologies.

References

Krizhevsky, Alex & Sutskever, Ilya & Hinton, Geoffrey. (2012). ImageNet Classification with Deep Convolutional Neural Networks. Neural Information Processing Systems. 25. https://doi.org/10.1145/3065386

]J. Wei et al., “Machine learning in materials science,” vol. 1, no. 3, pp. 338–358, 2019.

Agrawal A, Gans J, Goldfarb A. Economic policy for artificial intelligence.Innovation Policy and the Economy 2019;19:139–59.

Alqahtani, H., Kavakli-Thorne, M., & Kumar, G. (2021). Applications of generative adversarial networks (gans): An updated review. *Archives of Computational Methods in Engineering*, *28*, 525-552.

M. Kaur and A. Mohta, “A Review of Deep Learning with Recurrent Neural Network,” *2019 International Conference on Smart Systems and Inventive Technology (ICSSIT)*, Tirunelveli, India, 2019, pp. 460–465, https://doi.org/10.1109/ICSSIT46314.2019.8987837.

S. Samine, M. Zemzami, N. Hmina, M. Lagache and S. Belhouideg, "Towards the use of artificial intelligence and machine learning in material scientist field," 2022 8th International Conference on Optimization and Applications (ICOA), Genoa, Italy, 2022, pp. 1–5, https://doi.org/10.1109/ICOA55659.2022.9934559.

S. Samine, M. Zemzami, N. Hmina, M. Lagache and S. Belhouideg, "Using machine learning and deep learning in additive manufacturing," *2023 9th International Conference on Optimization and Applications (ICOA)*, AbuDhabi, United Arab Emirates, 2023, pp. 1–5, https://doi.org/10.1109/ICOA58279.2023.1030,8810.

S. Samine, M. Zemzami, N. Hmina, M. Lagache, and S. Belhouideg, Optimizing materials for 3D printing through Artificial Intelligence, 2024 International Conference on Connected Innovation and Technology (ICCITX), Rabat, Morocco, 2024, pp. 1–7. https://doi.org/10.1109/ICCITX61791.2024.11070508

Q. He, H. Zheng, X. Ma, L. Wang, H. Kong, and Z. Zhu, "Artificial intelligence application in a renewable energy-driven desalination system: A critical review," *Energy and AI*, vol. 7, p. 100123, 2022/01/01/ 2022.

Hossain, M. A., & Sajib, M. S. A. (2019). Classification of image using convolutional neural network (CNN). *Global Journal of Computer Science and Technology*, *19*(2), 13-14.

Kappes, B., Moorthy, S., Drake, D., Geerlings, H., and Stebner, A., 2018, "Machine Learning to Optimize Additive Manufacturing Parameters for Laser Powder Bed Fusion of Inconel 718," 9th International Symposium on Superalloy 718 & Derivatives: Energy; Aerospace, and Industrial Applications, Springer International Publishing, pp. 595–610.

Teja Kattenborn, Jens Leitloff, Felix Schiefer, Stefan Hinz,Review on Convolutional Neural Networks (CNN) in vegetation remote sensing,ISPRS Journal of Photogrammetry and Remote Sensing,Volume 173,2021,Pages 24-49, https://doi.org/10.1016/j.isprsjprs.2020.12.010.

Ketkar, N., Moolayil, J. (2021). Convolutional Neural Networks. In: Deep Learning with Python. Apress, Berkeley, CA. https://doi.org/10.1007/978-1-4842-5364-9_6.

Y. LeCun, Y. Bengio, and G. J. n. Hinton, "Deep learning," vol. 521, no. 7553, pp. 436–444, 2015.

LeCun, Yann & Bengio, Y. & Hinton, Geoffrey. (2015). Deep Learning. Nature. 521. 436-44. https://doi.org/10.1038/nature14539

Y. LeCun *et al.*, "Backpropagation applied to handwritten zip code recognition," vol. 1, no. 4, pp. 541–551, 1989.

S. Li, G. W. Hickey, M. M. Lander, and M. K. Kanwar, "Artificial Intelligence and Mechanical Circulatory Support," Heart Failure Clinics, vol. 18, no. 2, pp. 301–309, 2022/04/01/ 2022.

Liu Y, Zhao T, Ju W, Shi S. Materials discovery and design using machine learning. J Mater 2017;3:159e77.

Lopez-Jimenez F, Attia Z, Arruda-Olson AM, et al. Artificial intelligence in cardiology: present and future. Mayo Clin Proc 2020;95:1015-39.

K. Ma, S. A. Harmon, I. S. Klyuzhin, A. Rahmim, and B. Turkbey, "Clinical Application of Artificial Intelligence in Positron Emission Tomography: Imaging of Prostate Cancer," PET Clinics, vol. 17, no. 1, pp. 137–143, 2022/01/01/ 2022.

Mak HWL, Han R, Yin HHF. Application of Variational AutoEncoder (VAE) Model and Image Processing Approaches in Game Design. *Sensors*. 2023; 23(7):3457. https://doi.org/10.3390/s23073457.

Rachna Vaish, U.D. Dwivedi, Saurabh Tewari, S.M. Tripathi,Machine learning applications in power system fault diagnosis: Research advancements and perspectives,Engineering Applications of Artificial Intelligence,Volume 106,2021,104504, https://doi.org/10.1016/j.engappai.2021.104504.

Sahu, S.K.; Mokhade, A.; Bokde, N.D. An Overview of Machine Learning, Deep Learning, and Reinforcement Learning-Based Techniques in Quantitative Finance: Recent Progress and Challenges. *Appl. Sci.* **2023**, *13*, 1956. https://doi.org/10.3390/app13031956.

Sakurada, M., & Yairi, T. (2014). Anomaly Detection Using Autoencoders with Nonlinear Dimensionality Reduction. *Proceedings of the MLSDA 2014 2nd Workshop on Machine Learning for Sensory Data Analysis*, 4–11.

Jürgen Schmidhuber,Deep learning in neural networks: An overview,Neural Networks, Volume 61,2015, Pages 85–117,ISSN 0893–6080, https://doi.org/10.1016/j.neunet.2014.09.003.

A. M. Turing, "I.—COMPUTING MACHINERY AND INTELLIGENCE," *Mind*, vol. LIX, no. 236, pp. 433–460, 1950.

Wang, Quan. (2012). Kernel Principal Component Analysis and its Applications in Face Recognition and Active Shape Models.

Xu, H.; Chai, L.; Luo, Z.; Li, S. Stock movement prediction via gated recurrent unit network based on reinforcement learning with incorporated attention mechanisms. *Neurocomputing* **2022**, *467*, 214–228.

L. Yang, Z. Li, S. Ma, and X. Yang, "Artificial intelligence image recognition based on 5G deep learning edge algorithm of Digestive endoscopy on medical construction," *Alexandria Engineering Journal*, vol. 61, no. 3, pp. 1852–1863, 2022/03/01/ 2022.

C. Yin, Y. Zhu, J. Fei and X. He, "A Deep Learning Approach for Intrusion Detection Using Recurrent Neural Networks," in *IEEE Access*, vol. 5, pp. 21954-21961, 2017, https://doi.org/10.1109/ACCESS.2017.2762418.

Developing an Artificial Intelligence Model for Automatic Error Detection in 3D Prints Using Photos of Correct and Incorrect Prints

Sara Samine, Mohamed Karouchi, Maria Zemzami, Mohamed Gouskir, Nabil Hmina, Manuel Lagache, and Soufiane Belhouideg

Abstract

In the constantly evolving field of additive manufacturing, it is crucial to ensure the quality of 3D-printed parts to guarantee their reliability and performance. This paper presents a new technique for using a convolutional autoencoder to identify visual anomalies in 3D-printed components. The technique was used to understand the essential visual features of high-quality prints by training it on a dataset comprising flawless and defected photos. After training, the model was evaluated on an independent set of images, revealing considerable reconstruction errors in the samples with defects, but enabling defect-free parts to be reconstructed with few deviations. The produced anomaly maps showed regions with notable variations, suggesting possible flaws. The findings of this study open the door for better quality control procedures in additive manufacturing processes by showcasing the potential of convoluted autoencoders as an efficient automated tool for evaluating 3D quality.

Keywords

3D printing · Artificial Intelligence (AI) · Image analysis · Convolutional · Autoencoders · CNN

S. Samine (✉) · N. Hmina
Laboratory of Advanced Systems Engineering, National School of Applied Sciences ENSA, Ibn Tofail University Campus, Kenitra, Morocco
e-mail: sara.samine@uit.ac.ma

N. Hmina
e-mail: hmina@uit.ac.ma

M. Karouchi
Laboratory of Engineering in Chemistry and Physics of Matter Faculty of Sciences and Technics, Sultan Moulay Slimane University, Beni Mellal, Morocco
e-mail: Karouchimohamed3@gmail.com

M. Zemzami
National Graduate School of Arts and Crafts, Mohamed V University, Rabat, Morocco
e-mail: maria.zemzami@gmail.com; m.zemzami@um5r.ac.ma

M. Gouskir
National School of Applied Sciences, Sultan Moulay Slimane University, Béni Mellal, Morocco
e-mail: M.GOUSKIR@usms.ma

M. Lagache
Polytech Annecy Chambery, University of Savoie, SYMME, Annecy, France
e-mail: manuel.lagache@univ-savoie.fr

S. Belhouideg
Research Laboratory in Physics and Sciences for Engineers (LRPSI), Polydisciplinary Faculty, Sultan Moulay Slimane University, Béni-Mellal, Morocco
e-mail: soufianebelhouideg@usms.ma

1 Introduction

Additive manufacturing, or three-dimensional (3D) printing, has become a strategic production technology in many industrial sectors due to its ability to produce geometrically complex and customized parts on demand. Unlike subtractive manufacturing methods, where material is removed from a block to obtain the desired shape, 3D printing builds objects layer by layer using polymers, metals, or composites, offering unmatched flexibility (Gibson et al. 2015). Its applications span from aerospace to healthcare and the automotive sector, enabling the production of specialized components that meet rigorous design requirements while reducing both costs and production times.

However, the diversity of 3D printing technologies and materials introduces significant challenges in quality control. Each field of application requires specific printing configurations, making quality control more complex and prone to errors. For instance, in the fashion or jewelry industries, pieces

M. Gouskir et al. (eds.), *Cutting-Edge Artificial Intelligence Applications*, Sustainable Artificial Intelligence-Powered Applications,
https://doi.org/10.1007/978-3-032-00125-2_2

require precise finishes and refined details that are highly dependent on printing parameters (Yap and Yeong 2014). This variability generates uncertainties in the control of produced parts, as even slight variations in printing settings can lead to defects such as surface irregularities, dimensional errors, or weak inter-layer adhesion.

Traditional manual inspections, still widely used to detect such defects, are not only costly and time-consuming but also limited by human subjectivity. In large-scale production environments, they can result in expensive errors if defects go undetected (Leach et al. 2019). Furthermore, uncertainties in measurements and the metrology of 3D printing processes amplify the challenges of standardization, requiring more robust tools to ensure reliable quality control. Inaccuracies stemming from temperature settings, variations in material quality, or fluctuations in environmental conditions highlight the need for automated control methods capable of managing such variability.

In this context, advances in artificial intelligence (AI), particularly in machine learning and computer vision, offer an innovative solution to automate and enhance defect detection in 3D-printed parts. Traditional quality control methods in 3D printing, such as manual inspections and sensor-based monitoring, often have limitations in terms of accuracy, efficiency, and scalability. Manual inspections are time-consuming and prone to human errors, while sensor-based approaches require complex calibrations and often struggle to detect high-resolution defects. In contrast, AI-based methods, including convolutional neural networks (CNNs) and autoencoders, provide an automated and scalable solution capable of detecting defects with high precision.

Using machine learning algorithms, such as CNNs, defects can be detected automatically by analyzing images of printed parts, enabling rapid classification between correct and incorrect prints (Djenouri and G. Srivastava et JC-W. Lin, 2024). CNNs, as algorithms specialized in image recognition, have proven effective in detecting defects in printing images, even when such defects are not visible to the naked eye (Goodfellow et al. 2016). Thanks to their ability to extract visual features and recognize complex patterns, these networks can outperform traditional manual inspections in terms of both accuracy and speed. However, while CNNs excel in supervised classification tasks, they require large labeled datasets.

Numerous studies have applied CNNs for defect detection in manufacturing processes, including additive manufacturing. This variability generates uncertainties in the control of produced parts, as even slight variations in printing settings can lead to defects such as surface irregularities, dimensional errors, or weak inter-layer adhesion (Samine et al. 2025). For example, Cheng et al. (Cheng et al. 2019) demonstrated how CNNs can automatically detect surface irregularities, dimensional deviations, and other critical defects in 3D-printed parts by analyzing images captured at different stages of the manufacturing process. This type of intelligent system can quickly alert operators to defects, allowing real-time adjustments and improving the reliability of the produced parts. Additionally, Zhou et al. (Zhou et al. 2021) reviewed best practices and challenges in applying CNNs for defect detection, emphasizing the need for robust models capable of handling variations in defects and process parameters.

While CNNs are highly effective in supervised learning scenarios, autoencoders offer a compelling alternative for defect detection in unsupervised settings. Autoencoders identify anomalies by reconstructing images and detecting discrepancies between the original and generated outputs, making them particularly useful for defect detection in dynamic printing environments. This method enables the detection of subtle variations without the need for extensive labeled datasets, addressing one of the main challenges associated with supervised learning techniques.

Nevertheless, while AI-driven approaches show great promise for automating defect detection, it remains essential to consider the uncertainties and variations inherent in additive manufacturing processes. Measurement uncertainties, the diversity of defect types, and the complexity of printing configurations require AI models that can adapt to different production conditions to ensure reliable and generalizable defect detection (Djenouri and G. Srivastava et JC-W. Lin, 2024). By exploring the use of CNNs for the automatic classification of 3D prints, this article highlights not only the effectiveness of neural networks in identifying defects in 3D printing images but also the importance of accurate classification to enhance automated quality control in production.

2 Artificial Intelligence (AI) in 3D Printing and Quality Control

Artificial Intelligence (AI) has significantly impacted various fields, including 3D printing (Samine et al. 2022) (Samine et al. 2023), by offering innovative solutions to challenges in quality control and error detection. Through automated image analysis, AI can detect anomalies or defects without human intervention, enhancing both precision and efficiency in industrial processes.

Among the most effective techniques for image recognition and processing are Convolutional Neural Networks (CNNs) (Lee and Park xxxx). These networks excel at extracting complex features from images. Specifically designed for visual data, CNNs leverage the spatial structure of images to identify repetitive patterns, contours, textures, and other distinctive elements. This makes them particularly useful for detecting defects in 3D printing, where flaws are not always visible to the naked eye.

An advanced approach involves using CNNs as autoencoders, a form of unsupervised learning model (Mehta and Klarmann 2024). Unlike supervised models that require labeled data, autoencoders learn directly from unlabeled data. They work by compressing input images into a smaller representation and then reconstructing them from that compressed form.

An autoencoder consists of two key parts: an encoder, which reduces the original image, and a decoder, which tries to reconstruct it. The goal is to achieve an accurate reconstruction, and any discrepancy between the original and reconstructed image indicates an anomaly. These discrepancies often correspond to production errors in 3D printing, such as surface defects, geometric distortions, or missing layers. Such anomalies become apparent when the reconstructed image significantly differs from the original [13].

One of the main advantages of this method is that it eliminates the need for manually labeling potential defects, saving time and reducing the cost of error detection. Additionally, this approach can be integrated into automated production workflows, enabling continuous monitoring of 3D prints (Samine et al. 2024).

However, autoencoders have certain limitations when it comes to detecting anomalies. They may fail to identify defects that are visually close to normal variations, provide no explanation of the nature of the anomalies, and are sensitive to noise, which can generate false positives. Secondary classification could improve their reliability.

In summary, leveraging AI and CNNs as autoencoders provides an effective and efficient solution for defect detection in 3D printing. Automating this task ensures high-quality production while minimizing human intervention and the costs associated with errors.

3 Methodology

The methodology adopted in this study focuses on applying a convolutional autoencoder for anomaly detection in images of 3D-printed parts. The process is divided into several steps: data preparation, model design, model training, and anomaly detection.

3.1 Data Preparation

The images used in this study are photographs of 3D-printed parts, including both defective and non-defective objects. The data preparation process involves the following steps:

- Image acquisition: Images of 3D-printed parts are collected from various sources, including prints with potential defects.
- Preprocessing: Images are converted to grayscale to simplify processing and reduce computational complexity. They are then resized to a fixed resolution of 128 $\times$ 128 pixels. This resizing standardizes image dimensions and optimizes the model's learning process.
- Normalization: Each pixel in the image is normalized to a range of values between 0 and 1 by dividing the intensity by 255. This normalization speeds up learning by ensuring faster convergence of the model.

3.2 Autoencoder Model Design

The autoencoder used in this study consists of two main components: an encoder and a decoder. The encoder gradually reduces the spatial dimensions of the image while capturing essential features, while the decoder reconstructs the image from its compressed representation.

- Encoder: The encoder includes convolutional layers with (3 $\times$ 3) filters and ReLU activation functions. Each convolutional layer is followed by a max-pooling operation, which reduces the image size while retaining important features.
- Decoder: The decoder reconstructs the original image from its compressed form. It employs upsampling layers and convolutions to restore the original dimensions of the image. The final layer uses a sigmoid activation function to produce a reconstructed image with pixel values ranging from 0 to 1.

The model is implemented using the Keras library and optimized using the Adam algorithm, which is commonly used for convolutional neural networks.

3.3 Model Training

The autoencoder model is trained in an unsupervised manner, meaning that no specific labels indicating anomalies in the images are required. The training process includes the following steps:

- Dataset: The set of images is split into a training set and a validation set. The training set is used to adjust the model parameters, while the validation set is employed to monitor the model's performance over training epochs.
- Loss function: The chosen loss function is binary cross entropy, which is suitable for tasks involving binary or grayscale image reconstruction. This function measures

the divergence between the pixels of the input image and those of the reconstructed image.

- Training parameters: The model is trained for 50 epochs with a batch size of 8. A 10% validation split is applied to evaluate the model's performance during training and to prevent overfitting.

3.4 Anomaly Detection

Once the autoencoder is trained, it is applied to the test images to detect anomalies. The detection method relies on the difference between the original image and the reconstructed image.

- Reconstruction: Each test image is processed through the autoencoder, generating a reconstructed version of the image.
- Difference calculation: The pixel-by-pixel difference between the original and reconstructed images is computed. Significant differences indicate a potential anomaly in the image.
- Anomaly visualization: The differences between the original image and the reconstructed image are visualized using a heat map. Regions with significant discrepancies are highlighted in red, indicating potential defective areas in the 3D-printed part.

4 Results Discussion

4.1 Results

To identify and localize anomalies in 3D-printed parts, we implemented a convolutional autoencoder model trained to distinguish between the visual features of correct and defective parts. The autoencoder learns a compressed representation of each image and then attempts to reconstruct it. Any significant difference between an original image and its reconstruction is interpreted as a potential anomaly. To evaluate the effectiveness of this approach, images of the parts were first normalized and resized before being fed into the model. The pixel-by-pixel differences between the original and reconstructed images are quantified and visualized as heat maps to highlight areas of discrepancy. The results obtained allow for the detection of visual defects in 3D prints in an unsupervised manner, leveraging the model's ability to capture abnormal variations relative to expected visual features.

Following the implementation of the convolutional autoencoder, we present the visual results to demonstrate its effectiveness in identifying anomalies in 3D-printed parts.

Figure 1 shows a comparison between original images, their reconstructions, and the corresponding anomaly maps generated by the model. The heat maps highlight areas where significant differences were observed, suggesting potential defects in these regions.

- **Original versus Reconstructed Images**: For each analyzed part, the model accurately reconstructed images of defect-free parts, producing minimal discrepancies between the original and reconstructed images. This observation suggests that the autoencoder has successfully learned the key visual characteristics of high-quality 3D-printed parts.
- **Anomaly Detection**: In contrast, parts containing visible defects exhibited higher reconstruction errors. These discrepancies are visualized in the anomaly maps as high-contrast regions, particularly in areas where surface irregularities or structural inconsistencies are present. The pixel-level difference between the original and reconstructed images allowed the model to isolate these regions with a high degree of accuracy.

4.2 Accurate and Early Defect Detection

The autoencoder enables the detection of defects even when they are minor or just beginning to form. Due to its sensitivity to subtle differences between normal prints and those with errors, this method can identify issues as soon as they appear. This is highly advantageous in a production process, as it allows errors to be corrected early, reducing material and time costs associated with defective prints.

4.3 Ease of Implementation in Production Environments

Unlike other quality control methods that may require specific equipment or complex preparation steps, the autoencoder can be directly integrated into an existing workflow as an image analysis software. It only requires images of the printed object to operate, making it easy to incorporate into a production line without significant additional hardware. Images captured at various stages of the printing process can be analyzed in real time to ensure print quality.

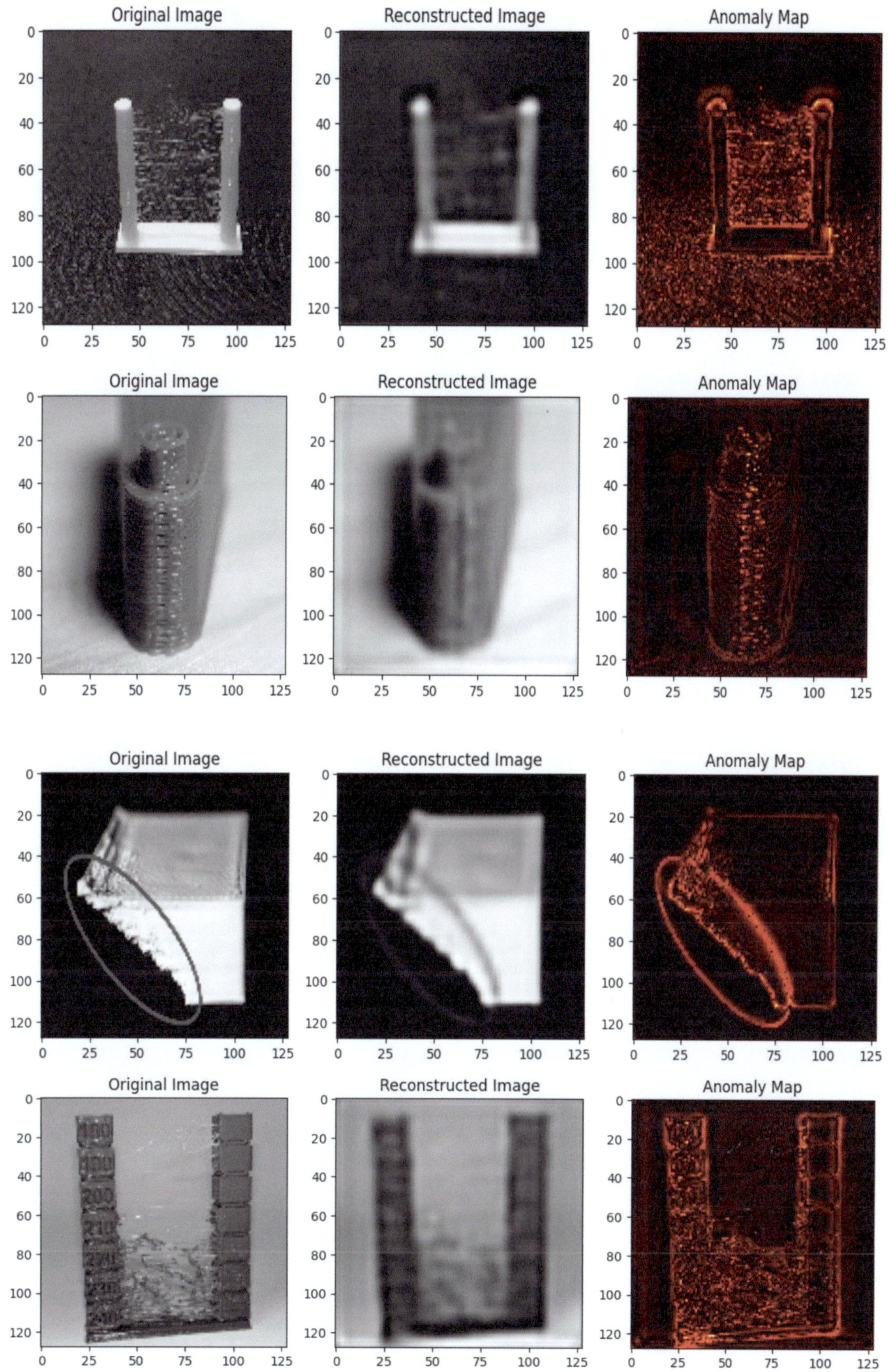

Fig. 1 Comparison of original images, reconstructed images, and anomaly maps for defect detection in 3D printing

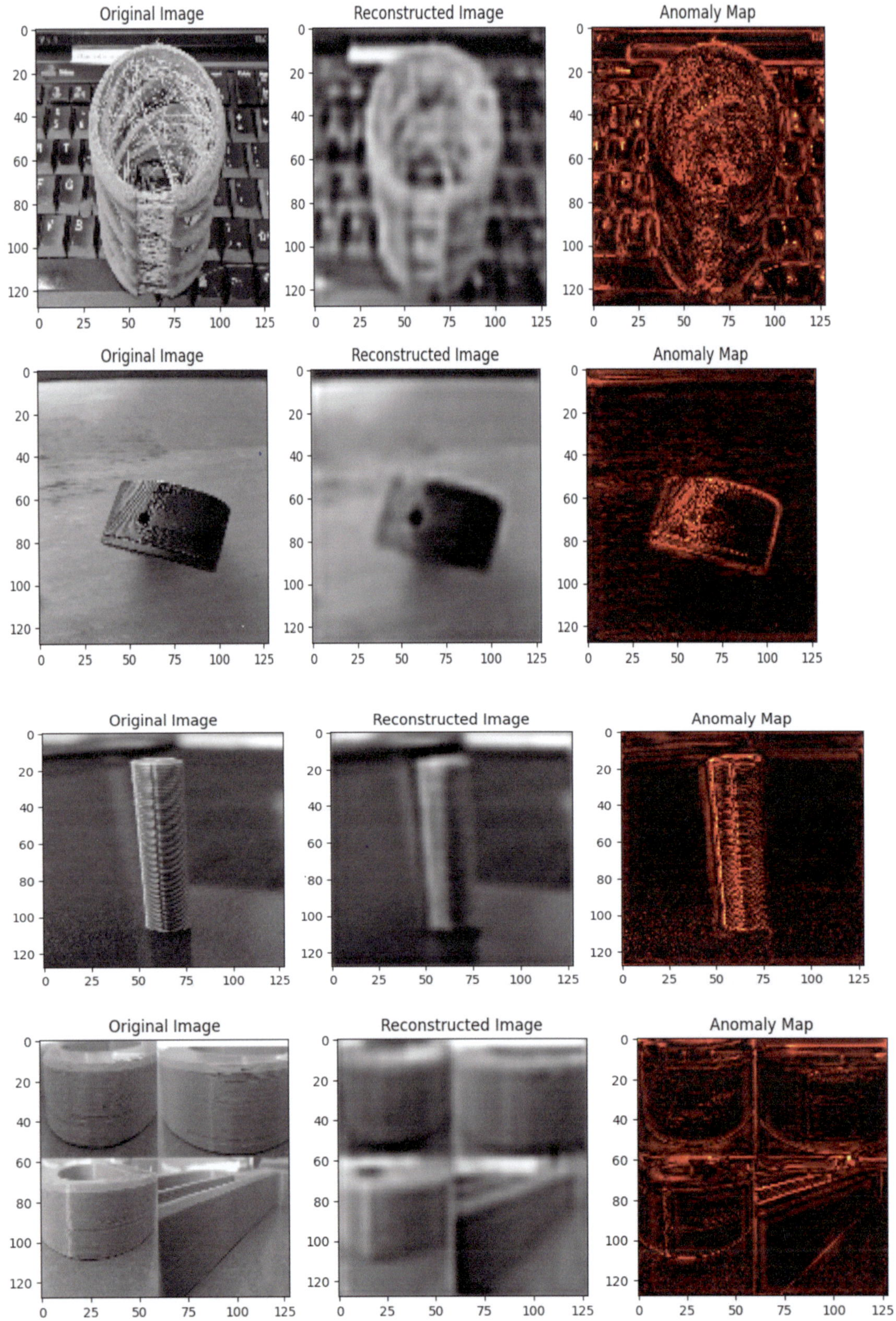

Fig. 1 (continued)

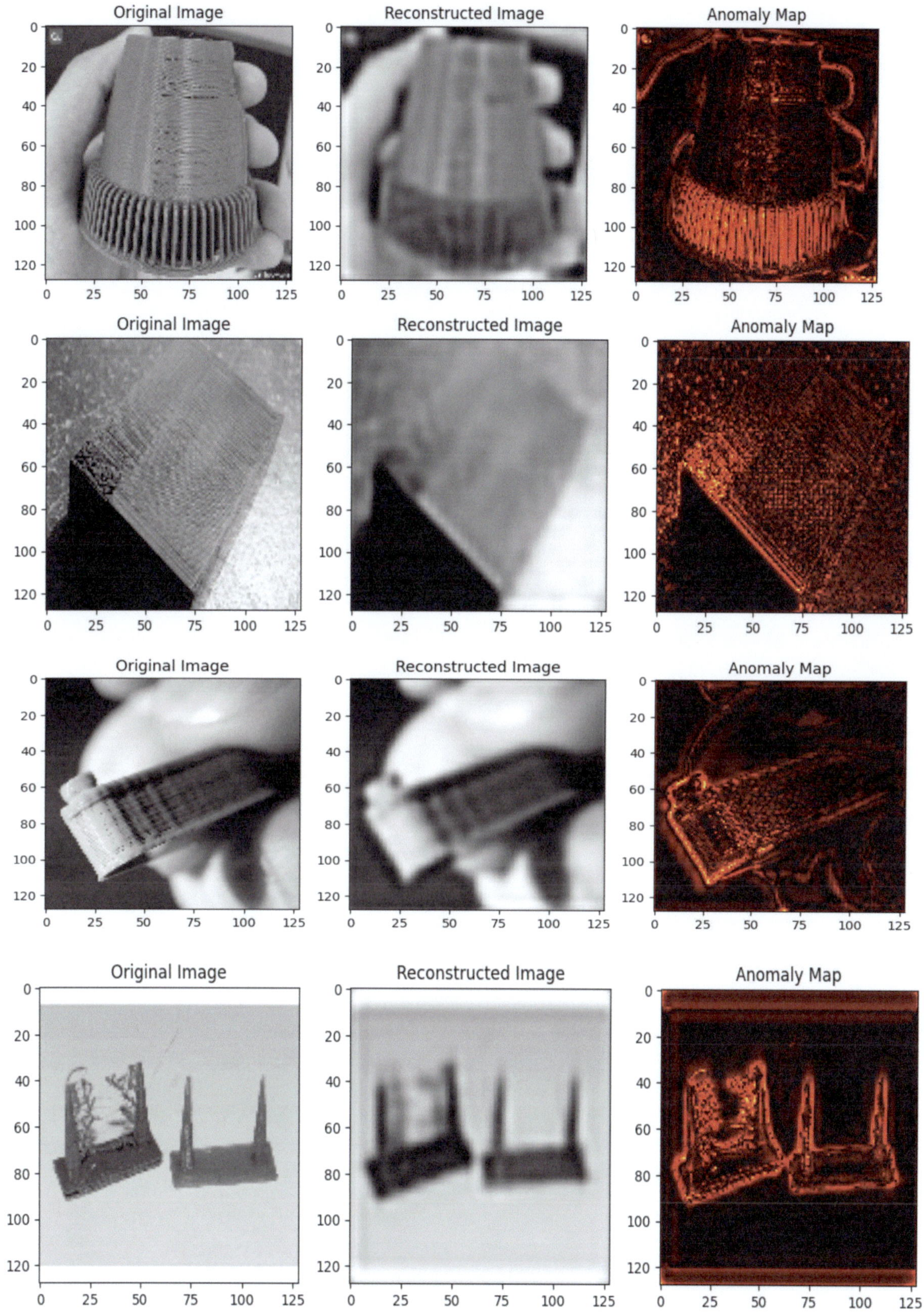

Fig. 1 (continued)

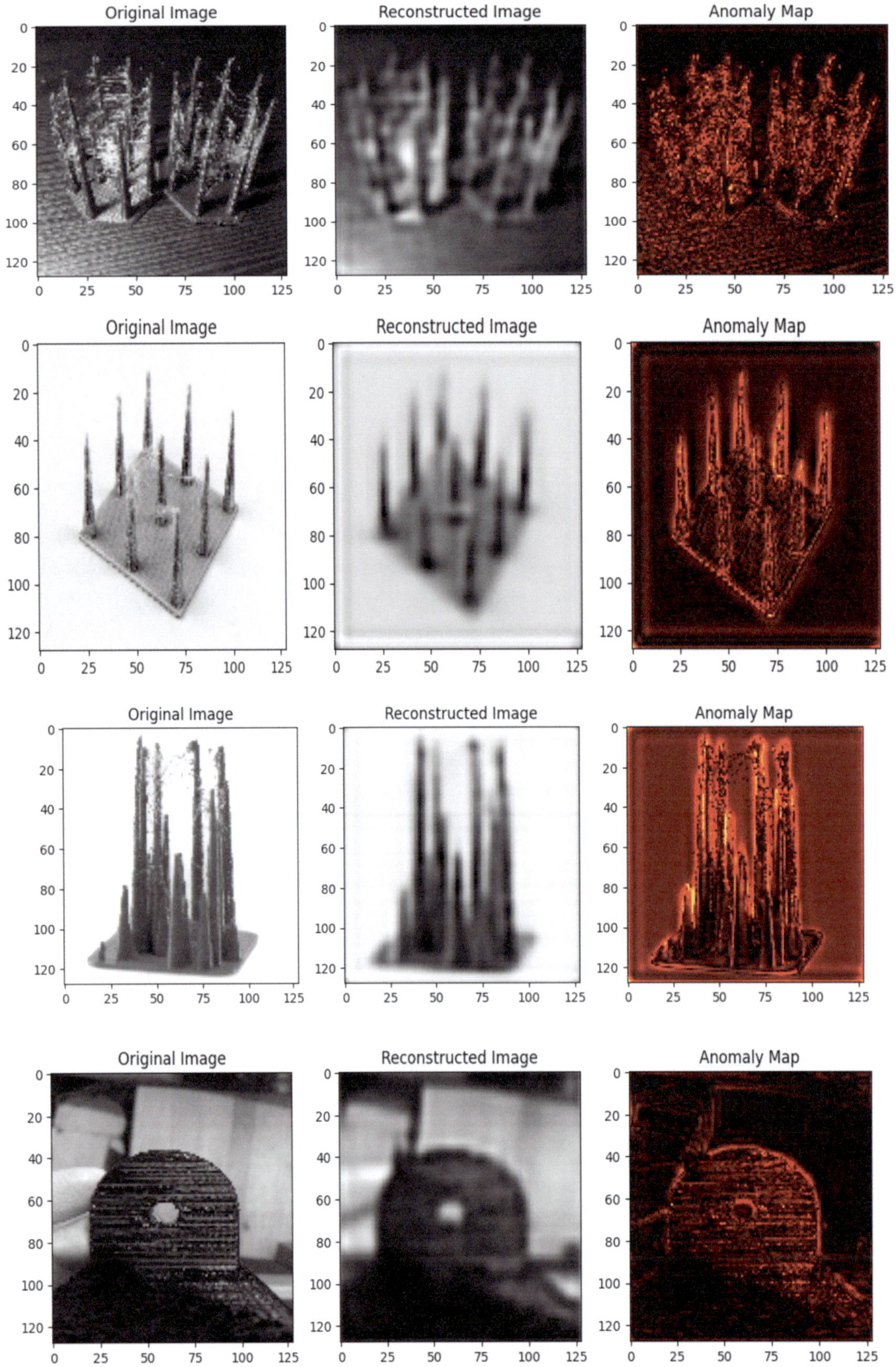

Fig. 1 (continued)

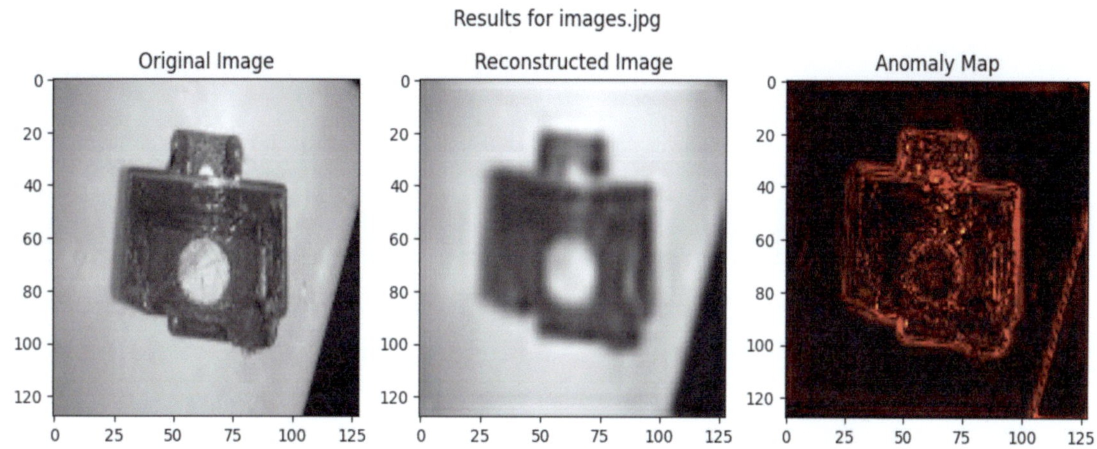

Fig. 1 (continued)

4.4 Adaptability and Generalization

This method effectively detects a wide range of defects, regardless of object geometry or surface complexity. The results show that the autoencoder can be applied to objects of varying shapes and maintains high sensitivity to anomalies. This means the method is flexible enough to be used on different types of parts and materials, making it a versatile and generalizable tool for diverse industrial environments.

4.5 Automation and Reduced Human Intervention

Implementing an anomaly detection model based on an autoencoder allows for the automation of quality control processes, reducing reliance on human inspection. This is particularly beneficial for batch printing or mass production, where manually inspecting each part would be impractical. With this method, anomalies are automatically identified and flagged, allowing operators to focus on problem-solving rather than detection.

4.6 Improvement of Product Quality and Reliability

By detecting and correcting errors quickly, this method contributes to the overall quality improvement of 3D-printed parts. It ensures that only parts meeting quality standards are retained, increasing customer satisfaction and enhancing production reliability. This can also boost the company's reputation for delivering high-quality finished products.

4.7 Resource Savings and Waste Reduction

By identifying defects before they propagate or affect large portions of the printed object, this method helps minimize raw material waste. This aligns with efforts to optimize resources and reduce costs by minimizing the need for reprinting or repairing parts.

4.8 Anomaly Detection in 3D Prints Using Autoencoder

This graph shows the evolution of the training and validation loss over 50 epochs during the autoencoder training. The loss is measured using binary cross entropy. We see a rapid decrease in loss during the initial epochs, indicating that the model is quickly learning to reconstruct the images (Figs. 2 and 3).

This graph displays the distribution of reconstruction errors for each tested image, where the mean reconstruction error represents the average difference between the original and reconstructed images. The majority of errors fall between 0.03 and 0.05, indicating that most images are well reconstructed by the autoencoder. Higher values (above 0.07) likely represent anomalies or poorly reconstructed images, which may suggest defects or errors in these specific 3D prints.

These graphs demonstrate that the autoencoder has successfully learned to reconstruct most of the 3D images and could be used as an anomaly detection model by identifying images with high reconstruction errors.

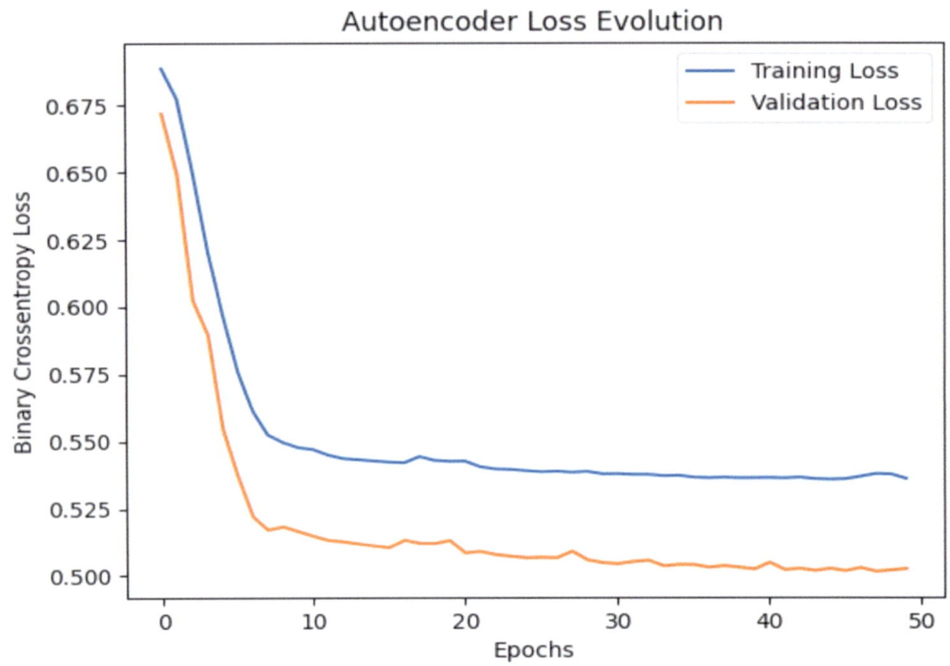

Fig. 2 Training and validation loss evolution during autoencoder training

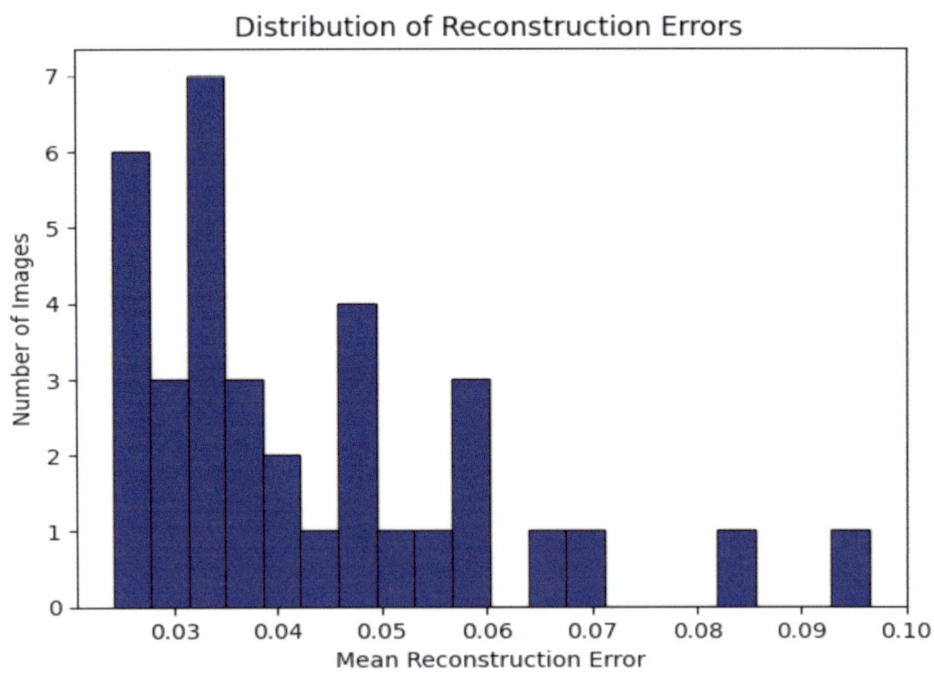

Fig. 3 Distribution of reconstruction errors across 3D-printed images

5 Conclusion

This study demonstrates the potential of using an autoencoder-based deep learning model for anomaly detection in 3D-printed images. By training the autoencoder on a dataset of 3D-printed samples, the model learns to reconstruct normal images accurately, enabling it to identify anomalies through higher reconstruction errors. The analysis of training and validation losses shows effective learning and convergence, while the distribution of reconstruction errors highlights the model's ability to distinguish between well-printed and defective parts.

Overall, this approach offers a promising, automated method for quality control in additive manufacturing, reducing the need for manual inspection and potentially enhancing the consistency and reliability of 3D printing processes.

References

Cheng, L., Jia, Z., & Yang, Z. (2019). *Detection of surface defects on 3D printed products based on deep learning.* Materials, 12(20), 3419. https://doi.org/10.3390/ma12203419

Y. Djenouri, G. Srivastava et JC-W. Lin, « Applied AI in Defect Detection for Additive Manufacturing: Current Literature, Metrics, Datasets, and Open Challenges », dans *IEEE Instrumentation & Measurement Magazine* , vol. 27, no. 4, pp. 46–53, juin 2024, https://doi.org/10.1109/MIM.2024.10540405.

S. Samine, M. Zemzami, N. Hmina, M. Lagache and S. Belhouideg, "Towards the use of artificial intelligence and machine learning in material scientist field," *2022 8th International Conference on Optimization and Applications (ICOA)*, Genoa, Italy, 2022, pp. 1–5, https://doi.org/10.1109/ICOA55659.2022.9934559.

S. Samine, M. Zemzami, N. Hmina, M. Lagache and S. Belhouideg, "Using machine learning and deep learning in additive manufacturing," *2023 9th International Conference on Optimization and Applications (ICOA)*, AbuDhabi, United Arab Emirates, 2023, pp. 1–5, https://doi.org/10.1109/ICOA58279.2023.10308810.

S. Samine, M. Zemzami, N. Hmina, M. Lagache, and S. Belhouideg, Optimizing Materials for 3D Printing through Artificial Intelligence, 2024 International Conference on Connected Innovation and Technology (ICCITX), Rabat, Morocco, 2024, pp. 1–7, https://doi.org/10.1109/ICCITX61791.2024.11070508.

Gibson, I., Rosen, D. W., & Stucker, B. (2015). *Additive manufacturing technologies: 3D printing, rapid prototyping, and direct digital manufacturing* (2nd ed.). Springer. https://doi.org/10.1007/978-1-4939-2113-3

Goodfellow, I., Bengio, Y., & Courville, A. (2016). *Deep learning.* MIT Press. https://www.deeplearningbook.org/

Leach, R. K., Bourell, D., Carmignato, S., Donmez, A., Senin, N., & Dewulf, W. (2019). *Geometrical metrology for metal additive manufacturing.* CIRP Annals, 68(2), 677-700. https://doi.org/10.1016/j.cirp.2019.05.004

Lee, Seung-mun and Park, Suk-Hee, Autonomous In-Situ Defect Detection and Correction in Additive-Lathe 3d Printing Process Using Variational Autoencoder Model. https://ssrn.com/abstract=4912815.

Mehta, D.; Klarmann, N. Autoencoder-Based Visual Anomaly Localization for Manufacturing Quality Control. *Mach. Learn. Knowl. Extr.* 2024, *6*, 1-17. https://doi.org/10.3390/make6010001.

Samine, S., Karouchi, M., Zemzami, M., Hmina, N., & Belhouideg, S. (2025). Mechanical properties of 3D-Printed aluminum, nickel, and titanium using a hybrid machine learning and computational mechanics approach. Frontiers in Materials, 12. https://www.frontiersin.org/journals/materials/articles/10.3389/fmats.2025.1584896, https://doi.org/10.3389/fmats.2025.1584896.

Simonyan, K.; Zisserman, A. Very Deep Convolutional Networks for Large-Scale Image Recognition. *arXiv* 2014. https://doi.org/10.48550/arXiv.1409.1556.

Yap, Y. L., & Yeong, W. Y. (2014). *Additive manufacture of fashion and jewellery products: A mini review.* Virtual and Physical Prototyping, 9(3), 195-201. https://doi.org/10.1080/17452759.2014.938993

Zhou, X., Chen, Y., Liu, Y., & Wang, L. (2021). *A survey on deep learning-based defect detection in additive manufacturing.* IEEE Transactions on Industrial Informatics, 17(5), 3364-3376. https://doi.org/10.1109/TII.2020.3014667

Effect of Wood Content and Orientation on the Micromechanical Behavior of 3D-Printed Wood-Polymer Composites

Ismail Ezzaraa, Nadir Ayrilmis, Mohamed Abouelmajd, Mouhcine Bakhaddache, Asmae Najm-Eddine, Youssef Najm-Eddine, Ahmed Bahlaoui, Ismail Arroub, Jamaa Bengourram, and Soufiane Belhouideg

Abstract

3D printing is revolutionizing the way parts are created, offering innovative solutions that compete with conventional manufacturing. In material extrusion, the introduction of filaments reinforced with natural fibers has considerably improved mechanical properties, making them useful in several sectors. However, before printing a part, it is essential to carry out numerical simulation to verify the impact of process parameters on the part quality. The aim of the present paper is to study the mechanical behavior of 3D-printed wood-polymer composites based on a numerical homogenization approach. The effects of wood volume fraction and orientation were discussed. The numerical results were verified using analytical models and experimental data, where a satisfactory agreement was obtained with the well-known Mori–Tanaka model.

Keywords

3D printing · Material extrusion · Strength · Stiffness · Homogenization · Micromechanics · Natural fiber · Polymer · Wood

I. Ezzaraa (✉) · J. Bengourram
Industrial Engineering Laboratory, Mechanical Engineering Department, Faculty of Science and Technology, University Sultan Moulay Slimane, Beni Mellal, Morocco
e-mail: ismail.ezzaraa@gmail.com

N. Ayrilmis
Department of Wood Mechanics and Technology, Forestry Faculty, Istanbul University-Cerrahpasa, Sariyer, Istanbul, Turkey

M. Abouelmajd · M. Bakhaddache · A. Najm-Eddine · Y. Najm-Eddine · A. Bahlaoui · I. Arroub · S. Belhouideg
Research Laboratory in Physics and Engineering Sciences, Team of Applied Physics and New Technologies, Department of Physics, Polydisciplinary Faculty, University Sultan Moulay Slimane, Beni Mellal, Morocco

1 Introduction

Strong, lightweight, and high-performance materials are in high demand for engineering applications. The weight reduction is a typical objective when designing parts with composite structures. Conventional methods such as injection molding, extrusion, and filament winding are often used to manufacture composite structures. However, they required several steps and are limited in their ability to manufacture different complex geometries. These technologies increase the cost of manufacturing lightweight parts.

Unlike traditional processes, 3D printing enables precise control of geometric shapes, internal structure, and surface through precise adjustment of the microstructure (Wang et al. 2020). This technology aims to print 3D objects by depositing material, layer by layer, until the complete object geometry is formed. It offers competitive advantages in terms of product design complexity, product personalization, and on-demand manufacturing. The emergence of 3D technology provides innovative strategies and new ways of developing novel systems.

Material extrusion (ME) is one of the most popular 3D printing methods. It involves passing the feedstock filament through a heated nozzle, which deposits the molten material in a controlled manner and in consecutive layers until the part is completed. Polylactic acid (PLA) is one of the materials commonly used in ME, due to its high strength, low thermal expansion, and low melting point, although it has low heat resistance unless annealed (Simmons et al. 2019). Therefore, to improve the properties of PLA and extend its lifespan, the material to be added to its structure must be of natural origin (Yilmaz et al. 2023).

3D-printed natural fiber composites (NFCs) are a relatively new group of eco-materials (Ezzaraa et al. 2023). After decades of high-tech development of synthetic fibers, such as carbon, aramid and glass, natural fibers like flax,

M. Gouskir et al. (eds.), *Cutting-Edge Artificial Intelligence Applications*, Sustainable Artificial Intelligence-Powered Applications,
https://doi.org/10.1007/978-3-032-00125-2_3

jute, and wood, have recently aroused great interset due to their biodegradable characteristics, low cost, low density, and comparable mechanical properties with synthetic fibers (Lila et al. 2018). As a result, many sectors are interested to these materials, particularly automotive, building construction, and furniture (Akemah and Ben-Alon 2023; Cheng et al. 2024; Deb and Jafferson 2021). Analysis of their mechanical properties is therefore essential for the use of these types of composites in advanced industries.

A great deal of research has been devoted to the development and characterization of new NFCs, in particular, 3D-printed wood/PLA composites (WPCs). For example, Estakhrianhaghighi et al. (2020) studied the thermo-mechanical properties of 3D-printed WPCs. The experimental results showed improved stiffness, ultimate strength, fracture strain, toughness, and thermal conductivity of reinforced samples, compared with the PLA. Ayrilmis et al. (2019) investigated the effects of wood flour (WF) content on the surface properties of 3D-printed materials produced from WF/PLA filaments, and found that incorporating WF into PLA filaments increased surface roughness and significantly decreased the wettability of the samples. Guo et al. (2018) fabricated WPCs using FDM and examined the effect of toughening agents on the properties of the printed specimens. The impact strength was increased by 51.31% when 10% of thermoplastic polyurethane was added as toughening agent. The mechanical properties of the samples and their composites with different additives (wood, ceramic, copper, aluminum, and carbon fiber) were evaluated by Liu et al. (2019), taking into account the effects of build orientations and raster angles. They found that PLA, wood-, and carbon fiber-based PLA exhibited better printing formability than the ceramic-, copper-, and aluminum-based PLA in upright orientation. Similar or even improved mechanical properties of ceramic-, copper- and aluminum-based PLA samples were achieved compared to pure PLA. Furthermore, the addition of wood and chopped carbon fiber to pure PLA considerably reduces its mechanical properties. In most cases, samples printed in on-edge orientation with +45°/−45° raster angles have the highest mechanical strength and modulus.

Despite the extensive investigations and development effort on 3D-printed WPCs for applications, most of them are based on experimental tests, which are time-consuming, costly, and complex. These limitations give rise to modeling and simulation tools based on micromechanical homogenization theory (Lei et al. 2023). This approach seems to be particularly suitable for composite materials, which are characterized by their multi-scale nature. The present paper attempts to evaluate the micromechanical properties of WPCs using numerical homogenization. The results obtained are confronted with experimental test and analytical results based on dilute model, Mori–Tanaka model, self-consistent model, etc. Parametric studies then highlight the influence of wood orientation and content on the mechanical properties of 3D-printed WPCs.

Table 1 Elastic properties of wood and PLA (Kariz et al. 2018)

Heading level	Young's modulus (GPa)	Poisson's ratio
PLA	3.72	0.35
Wood	14.31	0.4

2 Experimental Design

The feedstock used in this study was PLA Ingeo™ 2003D (supplied by NatureWorks, Blair, NE) in granulated form and wood fibers were obtained from a species of tree known as beech (Fagus sylvatica) (Kariz et al. 2018). The woodblock was ground in a laboratory mill (Retsch ZM 200), then sieved through a sieve, and only particles passing through an opening of 0.237 mm were used to produce the filaments.

Before preparing the mixtures, PLA granules and wood particles were dried in a laboratory dryer. Five different filaments were prepared with different wood volume fractions: 0%, 10%, 20%, 30%, 40%, and 50%. The elastic properties of wood and PLA are shown in Table 1.

After preparing the filaments, the 3D printing process begins with the design of the sample geometries. This modeling phase is carried out using computer-aided design software (SolidWorks), with dimensions conforming to ASTM D638 standard. A Zortrax M200 3D printer (Zortrax, Poland) was used to print the samples. The printer is based on FDM technology. The nozzle diameter was 0.4 mm and the print layer thickness 0.19 mm. The nozzle temperature was maintained at 230 °C and the printing speed was variable. For all filaments, samples were printed with dimensions of 40 mm × 12 mm × 4 mm (Fig. 1).

The tensile strength of the samples was measured using a universal testing machine (Zwick-Roel Z005). The test speed was 10 mm/min. Five samples were measured for each material and the mean values were determined.

3 Micromechanical Homogenization Theory

Micromechanical homogenization theory helps to evaluate the macroscopic phenomena (effective phenomena) on the basis of microscopic behavior, i.e., by taking into account the heterogeneities of the material (the volume fraction of their phases, their shape and orientation, etc.). The aim is to

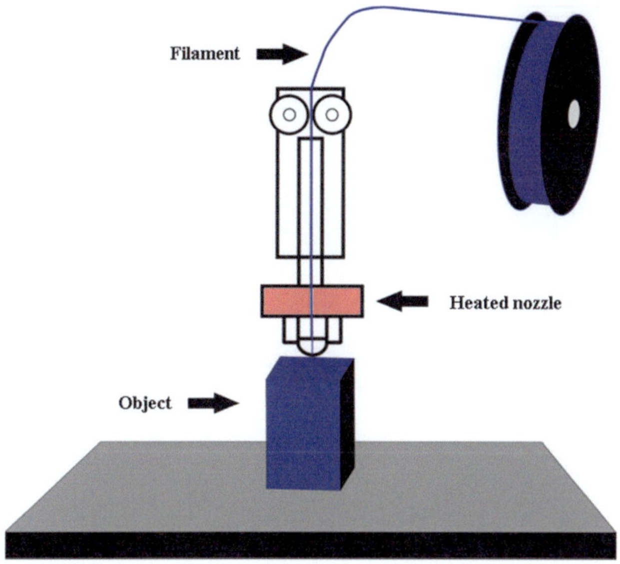

Fig. 1 Material extrusion principle

replace the heterogeneous material by an equivalent homogeneous material (EHM) with the same mechanical properties from a macroscopic point of view. Homogenization models consist of three steps: representation, localization, and homogenization.

- Representation: The step consists in defining the material. This involves defining the composite material and all the parameters to be taken into account.
- Localization: This step involves establishing a relationship between the microscopic strain/stress and the macroscopic strain/stress applied to the composite.
- Homogenization: This is the final stage, the aim to average and determine the actual behavior of the composite material.

Analytical and numerical homogenization methods have been widely developed and applied to a variety of composite materials. Analytical models, generally based on idealized representations, provide quick predictions or well-defined limits for the effective properties of composites. However, their accuracy remains insufficiently validated, particularly beyond certain specific microstructures (Ghossein and Lévesque 2012), which limits their use in industrial applications. Numerical methods, on the other hand, enable accurate calculation of effective properties by modeling a finite volume of the microstructure, making numerical homogenization a more reliable approach (Fig. 2).

In the numerical homogenization process, four steps are required to calculate the effective properties of a composite material, as illustrated in Fig. 3. The process begins with the generation of a 3D image of the random microstructure. This image is then discretized using the finite element analysis (FEA). Next, boundary conditions (BC) are applied to the model and, finally, the effective properties of the representative volume element (RVE) are calculated.

The constitutive law can be derived from the detailed field distributions within the selected RVE via an "averaging" process. Specifically, if the precise micro-fields σij and εij inside the RVE are known under the applied load, the average stresses and strains across the RVE can be expressed as follows:

$$\langle \sigma_{ij} \rangle = \frac{1}{V}\int_V \sigma_{ij}(x)dV \tag{1}$$

Fig. 2 Micromechanical homogenization process

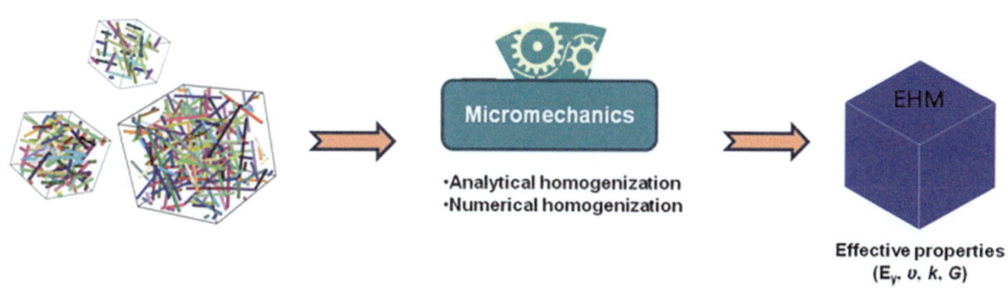

Fig. 3 Numerical homogenization process: **a**—microstructure generation; **b**—meshing; **c**—FE solution under BCs and; **d**—calculation of apparent properties

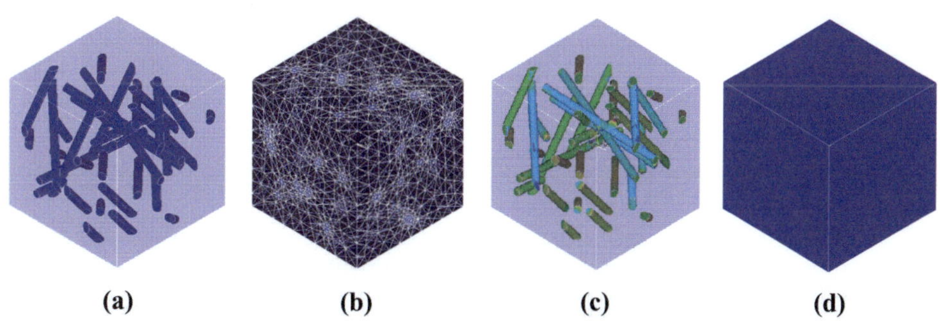

$$\langle \varepsilon_{ij} \rangle = \frac{1}{V}\int_{V} \varepsilon_{ij}(x)dV \tag{2}$$

Here, V represents the volume of the RVE. The averages values are taken as the effective stress and strain fields in the homogenized RVE. The relationship between σij and εij defines the "effective" constitutive law, often represented by Hooke's law:

$$\langle \sigma_{ij} \rangle = C_{ij}^{eff} : \langle \varepsilon_{ij} \rangle \tag{3}$$

Here, C_{ij}^{eff} denotes the components of the effective stiffness tensor. The number of independent constants within C_{ij}^{eff} is determined by the assumed symmetry of the material.

4 Results and Discussions

All finite element calculations were performed using DIGIMAT software, based on small displacement theory. Periodic boundary conditions were applied. For simplicity, the interface between the wood particles and the PLA matrix was assumed to be perfect. Furthermore, both materials are considered isotropic and exhibit linear elastic behavior. The particles were randomly distributed in the composite; their positions generated using the RSA (random sequential adsorption) algorithm.

4.1 Comparison Between Analytical and Numerical Results

The effective Young's modulus was evaluated for different wood volume fractions, ranging from 10 to 30%. To ensure the validity of our numerical model, comparisons were made with several analytical models, including the Voigt and Reuss bounds, the diluted model (DIL), the Mori–Tanaka model (MT), and the self-coherent model (SC). For a detailed explanation of these analytical models, readers can refer to the work of Ezzaraa et al. (2022). As shown in Fig. 4a, the incorporation of wood into the PLA matrix led to an increase in Young's modulus, thus improving the mechanical properties of the 3D-printed composite. This result was expected, because the

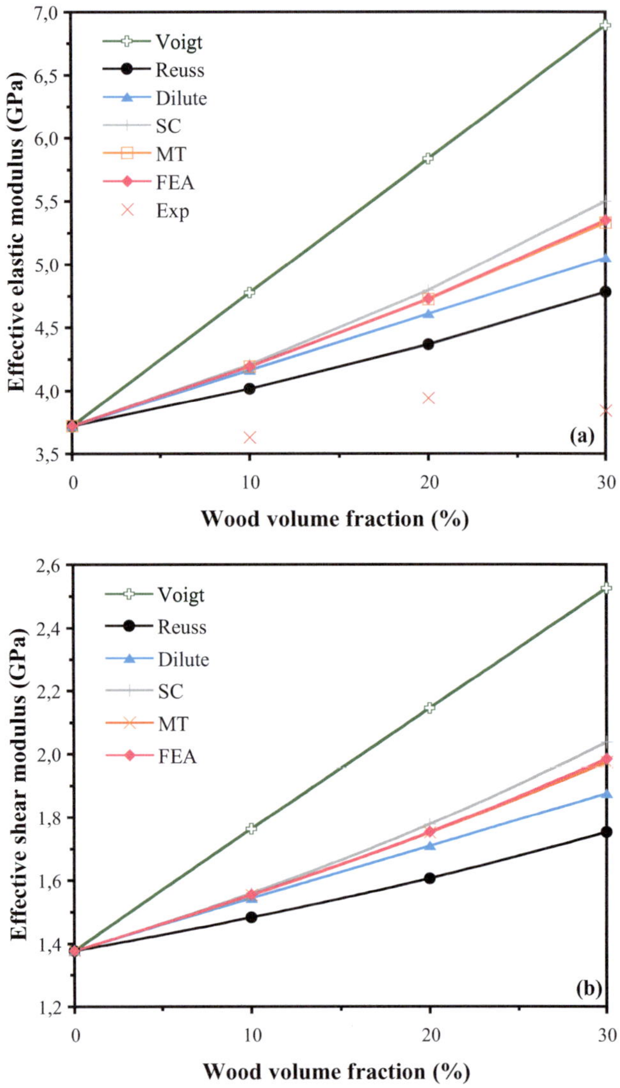

Fig. 4 Comparison of the predicted elastic properties using numerical and analytical approaches: **a**—elastic modulus, **b**—shear modulus

elastic modulus of wood is significantly higher than that of the PLA. Therefore, the addition of wood should increase the mechanical performances of the samples. Similar trends have been observed for the effective shear modulus, as illustrated in Fig. 4b.

Table 2 The components (GPa) of the stiffness tensors for aligned and random orientation

$$C_{aligned}^{eff} = \begin{bmatrix} 6.848 & 3.604 & 3.604 & 0 & 0 & 0 \\ 3.604 & 6.652 & 3.578 & 0 & 0 & 0 \\ 3.604 & 3.578 & 6.652 & 0 & 0 & 0 \\ 0 & 0 & 0 & 1.555 & 0 & 0 \\ 0 & 0 & 0 & 0 & 1.555 & 0 \\ 0 & 0 & 0 & 0 & 0 & 1.555 \end{bmatrix}$$

$\rightarrow \mathbf{E_{xx}} = 4.308\mathbf{GPa}, \mathbf{E_{yy}} = \mathbf{E_{zz}} = 4.16\mathbf{GPa}$

$$C_{random}^{eff} = \begin{bmatrix} 6.708 & 3.6 & 3.6 & 0 & 0 & 0 \\ 3.6 & 6.708 & 3.6 & 0 & 0 & 0 \\ 3.6 & 3.6 & 6.708 & 0 & 0 & 0 \\ 0 & 0 & 0 & 1.553 & 0 & 0 \\ 0 & 0 & 0 & 0 & 1.553 & 0 \\ 0 & 0 & 0 & 0 & 0 & 1.553 \end{bmatrix}$$

$\rightarrow \mathbf{E_{eff}} = 4.1933\mathbf{GPa}$

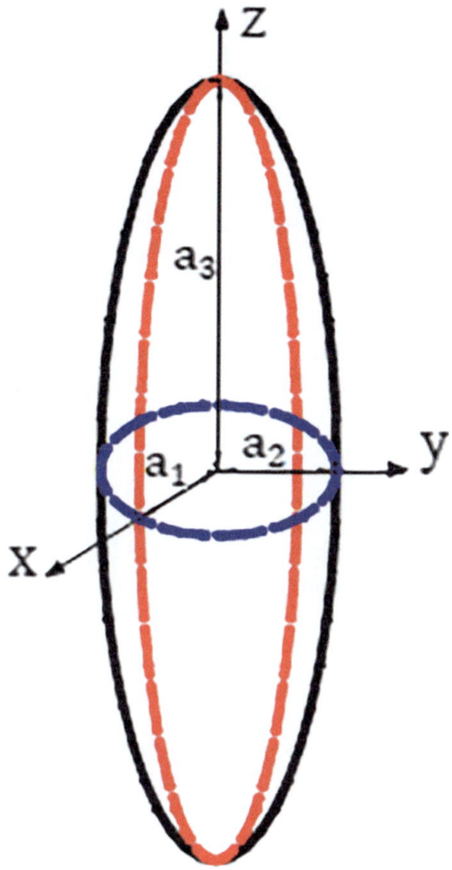

Fig. 5 Schematic illustration of an ellipsoidal particle

The figure also showed that for low volume fractions (<15%), the predictions of various analytical models closely align with those of FEA. However, as the volume fraction increases, noticeable differences begin to appear. The relative error in the elastic modulus prediction was 5.04% for the DIL model, 3.38% for the SC model, and only 0.19% for the MT model when compared to FEA, suggesting that the MT model provides results very similar to FEA.

Additionally, the experimental values were found to underestimate the FEA results. This can be attributed to the fact that the existing porosity in the material was not considered.

4.2 Effect of Wood Orientation

In composite materials, reinforcement orientation can significantly affect the part's mechanical behavior. The two directions studied are unidirectional (aligned along x-axis) and random directions (Fig. 6). Figure 7 represents predicted elastic properties E_{xx} (along the x-axis) and E_{yy} (along the y-axis) as a function of wood volume fraction for aligned and random wood orientation. For better clarity, Table 2 lists the components of the stiffness tensors for the two studied cases. The calculations are carried out for ellipsoidal particles ($a_1 = 0.23$ mm, $a_2 = a_3 = 0.11$ mm), and with a wood volume fraction of 10% (Fig. 5).

It can be seen that the elastic properties and anisotropy are strongly influenced by particles orientation. The equivalent material is isotopic in the case of random particles' orientation and transversely isotropic for aligned particles. This is expected since the particles are randomly distributed inside the matrix, leading to equal properties. While for composites reinforced with particles placed in a specific direction improve the mechanical properties in that direction. Thus, the composite behaves as a transversely isotropic material. Moreover, samples reinforced with aligned particles possess the

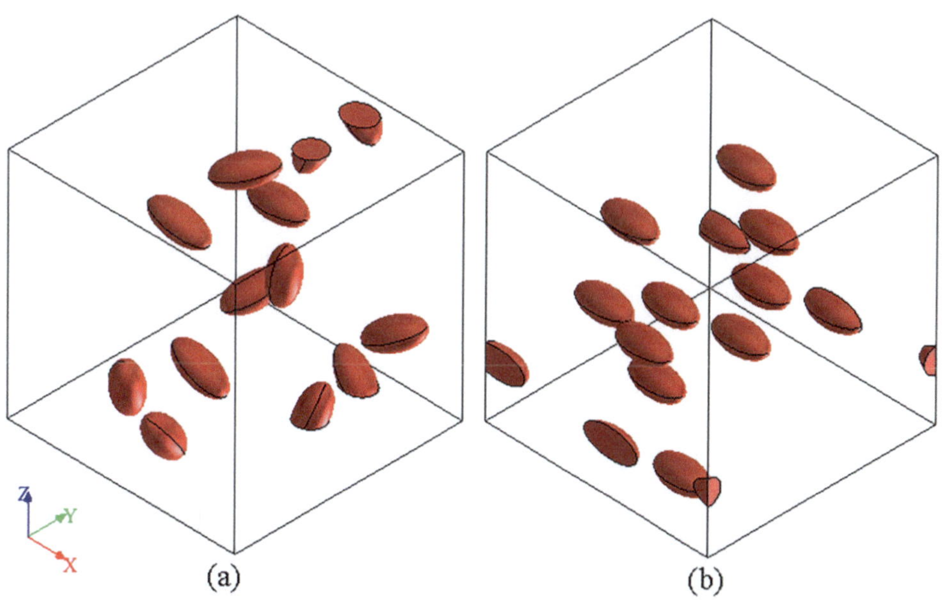

Fig. 6 RVE models with different fibers orientation: **a** random, **b** aligned

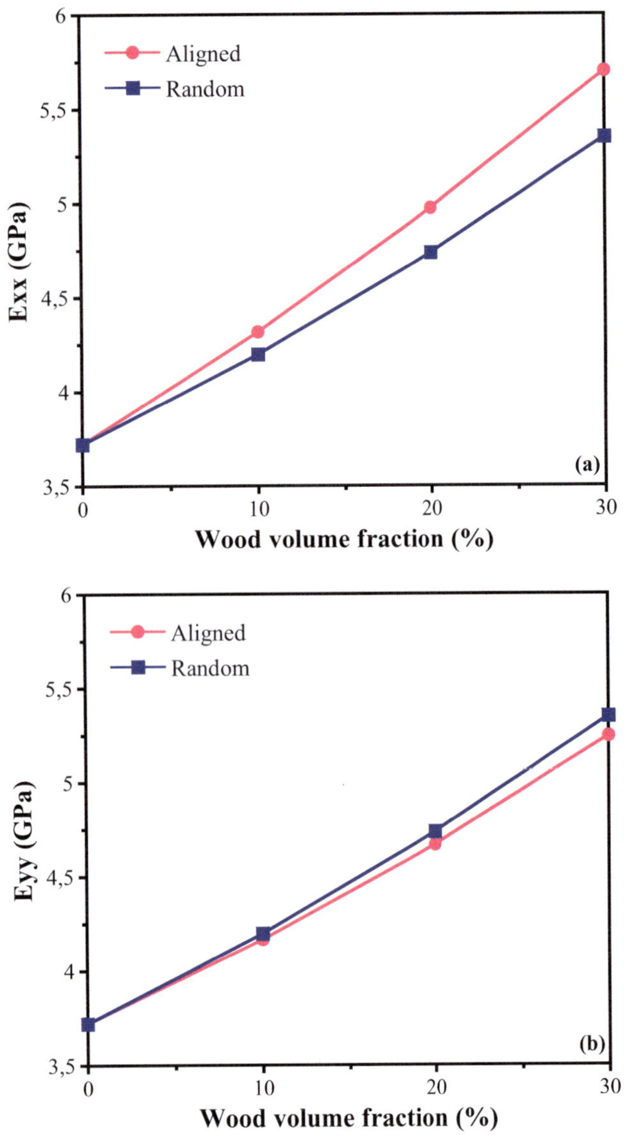

Fig. 7 Predicted Young's moduli for aligned and random wood orientation: **a**—elastic modulus along x-axis (E_{xx}), **b**—elastic modulus along y-axis (E_{yy})

higher elastic modulus (E_{xx}) compared to the random orientation. For example, E_{xx} was enhanced by 2.75% when passing from the case of random particles to aligned particles. On the other hand, the elastic modulus along the y-axis (E_{yy}) was almost the same for all cases.

5 Conclusion

The applications of 3D printing are constantly expanding, bridging the gap between the conceptualization of ideas and the development of products. Beyond polymers, researchers have developed 3D-printed WPCs, with ongoing studies to evaluate their properties. This article focuses on the examination of the impact of two specific parameters on the effective elastic properties of 3D-printed WPCs. The main conclusions are summarized as follows:

- The mechanical properties of the 3D-printed WPCs are significantly influenced by the wood fiber content and orientation.
- The importance of the fiber orientation on the anisotropy in the 3D-printed composite.
- A strong correlation was observed between the results of the FEA and the MT model, while discrepancies were noted with experimental data, likely due to the omission of the porosity phase in the analysis.

The results provide valuable insights for industrial applications such as automotive and building materials, where lightweight yet mechanically robust materials are essential. The ability to tailor fiber orientation and content can improve strength-to-weight ratios, making these composites suitable for interior automotive parts, structural panels, or eco-friendly building elements. Further investigations into porosity effect could help refine predictive models and expand the applicability of 3D-printed WPCs in these industries.

References

T. Akemah and L. Ben-Alon. Developing 3D-Printed Natural Fiber-Based Mixtures. in Bio-Based Building Materials. 2023. Cham: Springer Nature Switzerland.

N. Ayrilmis, M. Kariž, and M. Kitek Kuzman, Effect of wood flour content on surface properties of 3D printed materials produced from wood flour/PLA filament. International Journal of Polymer Analysis and Characterization, 24(7): p. 659–666 (2019)

P. Cheng, Y. Peng, K. Wang, et al., 3D printing continuous natural fiber reinforced polymer composites: A review. Polymers for Advanced Technologies, 35(1): p. e6242 (2024)

D. Deb and J.M. Jafferson, Natural fibers reinforced FDM 3D printing filaments. Materials Today: Proceedings, 46: p. 1308-1318 (2021)

E. Estakhrianhaghighi, A. Mirabolghasemi, Y. Zhang, et al., 3D-Printed Wood-Fiber Reinforced Architected Cellular Composites. Advanced Engineering Materials, 22(11): p. 2000565 (2020)

I. Ezzaraa, N. Ayrilmis, M.K. Kuzman, et al., Micromechanical models for predicting the mechanical properties of 3D-printed wood/PLA composite materials: A comparison with experimental data. Mechanics of Advanced Materials and Structures, 29(27): p. 6755-6767 (2022)

I. Ezzaraa, N. Ayrilmis, M. Abouelmajd, et al. Numerical Modeling Based on Finite Element Analysis of 3D-Printed Wood-Polylactic Acid Composites: A Comparison with Experimental Data. Forests, 14, (2023)

E. Ghossein and M. Lévesque, A fully automated numerical tool for a comprehensive validation of homogenization models and its application to spherical particles reinforced composites. International Journal of Solids and Structures, 49(11): p. 1387-1398 (2012)

R. Guo, Z. Ren, H. Bi, et al., Effect of toughening agents on the properties of poplar wood flour/poly (lactic acid) composites fabricated with Fused Deposition Modeling. European Polymer Journal, 107: p. 34-45 (2018)

M. Kariz, M. Sernek, M. Obućina, et al., Effect of wood content in FDM filament on properties of 3D printed parts. Materials Today Communications, 14: p. 135-140 (2018)

M. Lei, Y. Wang, Q. Wei, et al., Micromechanical modeling and numerical homogenization calculation of effective stiffness of 3D printing PLA/CF composites. Journal of Manufacturing Processes, 102: p. 37-49 (2023)

M.K. Lila, A. Singhal, S.S. Banwait, et al., A recyclability study of bagasse fiber reinforced polypropylene composites. Polymer Degradation and Stability, 152: p. 272-279 (2018)

Z. Liu, Q. Lei, and S. Xing, Mechanical characteristics of wood, ceramic, metal and carbon fiber-based PLA composites fabricated by FDM. Journal of Materials Research and Technology, 8(5): p. 3741-3751 (2019)

H. Simmons, P. Tiwary, J.E. Colwell, et al., Improvements in the crystallinity and mechanical properties of PLA by nucleation and annealing. Polymer Degradation and Stability, 166: p. 248-257 (2019)

H. Wang, Z. Deng, J. Chen, et al., A novel vehicle-like drug delivery 3D printing scaffold and its applications for a rat femoral bone repairing in vitro and in vivo. International Journal of Biological Sciences, 16(11): p. 1821-1832 (2020)

S. Yilmaz, O. Gul, B. Eyri, et al., Comprehensive characterization of 3D-printed bamboo/poly(lactic acid) bio composites. Polymer Engineering & Science, 63(9): p. 2958-2972 (2023)

Optimized High-Order Dual Hahn Moments for Image and Signal Reconstruction

Abdelati Bourzik, Belaid Bouikhalene, Jaouad El-Mekkaoui and Amal Hjouji

Abstract

This paper introduces a novel approach for automatically selecting the parameters of discrete Dual Hahn polynomials using a swarm optimization algorithm. The proposed method identifies the optimal parameters for constructing Dual Hahn moments, thereby improving the accuracy of image moments. Our approach minimizes the mean square reconstruction error, leading to significant improvements in image and signal reconstruction quality. The results demonstrate that the proposed method achieves lower reconstruction errors, especially at lower orders of dual Hahn moments. Additionally, the optimized dual Hahn moments were evaluated and compared against other parametric discrete moments, both with and without parameter optimization. The experiments clearly highlight the advantages of this approach, particularly at lower moment orders.

Keywords

Dual Hahn polynomials • Optimization • Image reconstruction • Signal reconstruction

1 Introduction

Discrete orthogonal moments have figured in many applications in computer version, including pattern recognition, image processing, and signal processing. Discrete moments such as Krawtchouk moments Asli and Flusser [2014], Bourzik et al. [2024], Tchebichef moments Bourzik et al. [2023], Charlier moments Bourzik et al. [2024], Hahn moments Zhou et al. [2005], Gegenbauer moments Hjouji et al. [2021], and Meixner moments Raj [2020]. All these moments have specified on a uniform lattice. Despite this, other discrete moments are orthogonal on a non-uniform lattice, for example, dual Hahn moments Zhu et al. [2007a] and Racah Zhu et al. [2007b]. However, statistical physics, quantum mechanics, and numerical analysis all heavily rely on discrete orthogonal moments with non-uniform lattice Papakostas et al. [2008]. A study using polynomials defined on a non-uniform lattice for 3D pattern recognition was published in Mesbah et al. [2017]. Later, in El Mallahi et al. [2018], the authors suggested new sets of rotation invariants for 2D and 3D based on orthogonal radial dual Hahn moments that are based on polar representation.

The Orthogonal Dual Hahn moments (DHMs) represent a significant advancement in the field of image and signal processing. They offer an innovative approach to analyzing images and signals by leveraging the properties of dual Hahn polynomials (DHPs), which are recognized for their discrete orthogonality. In the literature there is some works on the optimization of orthogonal moments like Charlier, Krawtchouk Bourzik et al. [2024], Meixner Bencherqui et al. [2022], Hahn moments Bourzik et al. [2024], and Jacobi moments Yahya et al. [2024]. According to our knowledge there is no such study for DHMs in low or high orders.

There are many optimization algorithms, here the Firefly Algorithm (FA) is used, which is a nature-inspired optimization technique based on the flashing behavior of fireflies. It is a type of swarm intelligence algorithm, due to its successful application across several fields according to its simplicity and effectiveness in solving complex optimization problems. Firefly algorithm can be applied in image processing and computer vision Dey et al. [2020]. It can help improve segmentation accuracy Sharma and Sehgal [2016], refine edge detection, and enhance feature extraction, leading to more robust

A. Bourzik (✉) · B. Bouikhalene
LIMATI Laboratory, Polydisciplinary Faculty, Sultan Moulay Slimane University, Beni Mellal, Morocco
e-mail: abdelati.bourzik@usms.ac.ma

J. El-Mekkaoui
LSEM Laboratory, Higher School of Technology, Sidi Mohamed Ben Abdellah University, Fez, Morocco

A. Hjouji
STIC Laboratory, Faculty of Science, Sidi Mohamed Ben Abdellah University, Fez, Morocco

M. Gouskir et al. (eds.), *Cutting-Edge Artificial Intelligence Applications*, Sustainable Artificial Intelligence-Powered Applications,
https://doi.org/10.1007/978-3-032-00125-2_4

and efficient visual data analysis. In this paper, we will elaborate a new way for automatically selecting the appropriate parameters of Dual Hahn polynomials for optimal signal and image reconstruction. Based on minimizing the reconstruction error, the swarm optimization technique returns the best values of DHP parameters. We apply this algorithm for high order computed by recurrence relations presented in the paper of Daoui et al. [2022]. Our method gives better results in low orders of dual Hahn moments.

The current paper is organized as follows. The second section presents a background on the discrete dual Hahn polynomials (DHPs). Followed in third section by the stable computation of DHPs and computation of Dual Hahn moments (DHMs). Section 4 is about the proposed method of DHM optimization and their parameter selection. Then, some experiment analysis to test the performances and comparison with manually selected parameters and discussions in Sect. 5. Finally, the paper finished with a conclusion.

2 Discrete Dual Hahn Polynomials

The dual Hahn polynomials of order n and discrete variable x are given by the following hypergeometric function according to Zhu et al. [2007a]:

$$h_n^{(a,b,c)}(x) = \frac{(a-b+1)_n(a+c+1)_n}{n!}\,{}_3F_2(-n, a-x, a+x+1; a-b+1, a+c+1; 1), \quad (1)$$

where $n = 1, 2, \ldots, N$, and $x = a, a+1, \ldots, b$. Here the parameters a, b, and c are restricted in Zhu et al. [2007a] to $:-\frac{1}{2} < a < b, |c| < a+1, b = a+N$.

And ${}_3F_2(.)$ is the hypergeometric function defined by the following relation:

$${}_3F_2(a_1, a_2, a_3; b_1, b_2; x) = \sum_{i=0}^{+\infty} \frac{(a_1)_k(a_2)_k(a_3)_k}{(b_1)_k(b_2)_k}\frac{x^k}{k!} \quad (2)$$

with $(a)_k$ is the Pochhammer symbol given by

$$(a)_k = a(a+1)(a+2)\ldots(a+k-1) = \frac{\Gamma(a+k)}{\Gamma(a)}. \quad (3)$$

$\Gamma(.)$ is the gamma function such that $\Gamma(x+1) = x!$

The orthogonality condition can be driven by the following scalar product:

$$\sum_{x=a}^{b} h_n^{(a,b,c)}(x)h_m^{(a,b,c)}(x)\rho(x)[\Delta\chi(x-\frac{1}{2})] = \delta_{nm}d_n^2, n, m = 1, 2, \ldots, N, \quad (4)$$

where $\Delta\chi(x-\frac{1}{2}) = 2x+1 > 0$ for $x > -\frac{1}{2}$ [Daoui et al., 2022], and δ_{pq} is the symbol of Kronecker such that

$$\delta_{nm} = \begin{cases} 1 \;\; if \;\; n = m; \\ 0 \;\; elsewhere. \end{cases} \quad (5)$$

The weighting functions $\rho(x)$ and d_n^2 are given by the following two equations Zhu et al. [2007a] in Table 1: The Dual Hahn Polynomials (DHPs) could be scaled by using the square norm and the weighting function. The weighted and scaled DHPs are defined by the following relation:

$$\widehat{h}_n^{(a,b,c)}(x) = h_n^{(a,b,c)}(x)\sqrt{\frac{\rho(x)}{d_n^2}[\Delta\chi(x-\frac{1}{2})]}. \quad (6)$$

So the weighted DHPs are thus confined to be orthogonal within the range of $[-1, 1]$, and the scalar product becomes

$$\sum_{x=a}^{b} \widehat{h}_n^{(a,b,c)}(x)\widehat{h}_m^{(a,b,c)}(x) = \delta_{nm}, n, m = 1, 2, \ldots, N. \quad (7)$$

3 Stable Computation of Dual Hahn Polynomials

The current polynomials in the previous section are not stable when the order of moments pass 80, which means that it is hard to deal with images more than 80×80 pixels.

The direct method for computation of dual Hahn polynomials suffers from instability for orders more than 80, for this reason a proposed method is presented to overcome this problem in Daoui et al. [2022]. Here, the first two orders of DHPs: $\widehat{h}_0^{(a,b,c)}(s)$ and $\widehat{h}_1^{(a,b,c)}(s)$ are calculated in such a way so as to simplify the terms that depend on d_n^2 and $\rho(s)$ to avoid numerical fluctuations in values Daoui et al. [2022]. Indeed,

- for $n = 0$ we have

$$\widehat{h}_0^{(a,b,c)}(s) = \sqrt{\frac{(a+s)(c+s)(b-s)(2s+1)}{(s-a)(s+b)(s-c)(2s-1)}}\widehat{h}_0^{(a,b,c)}(s-1), \quad (8)$$

where

$$\widehat{h}_0^{(a,b,c)}(a) = \sqrt{2a+1} * e^{\frac{E-F}{2}}, \quad (9)$$

with $E = ln\Gamma(2a+1) + ln\Gamma(b-c)$ and $F = ln\Gamma(a+b+1) + ln\Gamma(a-c+1)$, where $e^{\cdot}$ being the exponential function, and $ln\Gamma(.)$ being the logarithmic gamma function;

- for $n = 1$ we have

$$\widehat{h}_1^{(a,b,c)}(s) = G \times \frac{(s-a)(s+b)(s-c) + (s+a+1)(s+c+1)(s-b+1)}{(2s+1)}\widehat{h}_0^{(a,b,c)}(s), \quad (10)$$

where $G = \sqrt{\frac{1}{(a+c+1)(b-c-1)(b-a-1)}}$.

Table 1 Data for dual Hahn polynomials

$\rho(x)$	$\dfrac{\Gamma(a+x+1)\Gamma(c+x+1)}{\Gamma(x-a+1)\Gamma(b-x)\Gamma(b+x+1)\Gamma(x-c+1)}$
d_n^2	$\dfrac{\Gamma(a+x+p+1)}{n!(b-a-p-1)\Gamma(b-c-n)}$
A	$\dfrac{1}{n}[s(s+1)-ab+ac-bc-(b-c-a-1)(2n-1)+2(n-1)^2]$
B	$\dfrac{n}{(n+a+c)(b-a-n)(b-c-n)}$
C	$-\dfrac{1}{n}(a+c+n-1)(b-a-n+1)(b-c-n+1)$
D	$\dfrac{n(n-1)}{(n+a+c)(a+c+n-1)(b-a-n+1)(b-c-n+1)(b-a-n)(b-c-n)} = B \times \dfrac{1-n}{n \times C}$

The three-term recurrence equation below, which is a recursive technique with respect to the polynomial order n, is used to calculate $\widehat{h}_n^{(a,b,c)}(s)$ for orders $n > 1$ Zhu et al. [2007a]:

$$\widehat{h}_n^{(a,b,c)}(s) = A\sqrt{B}\widehat{h}_{n-1}^{(a,b,c)}(s) + C\sqrt{D}\widehat{h}_{n-2}^{(a,b,c)}(s), \qquad (11)$$

where A, B, C, and D are given in Table 1.

The stability criterion specified in the following equation is required for computing high-order DHPs in order to overcome unstable coefficients Daoui et al. [2022].

$$\widehat{h}_n^{(a,b,c)}(i) = \begin{cases} 0 \text{ if } (n > \frac{N}{2})\&(\mid \widehat{h}_n^{(a,b,c)}(i) \mid > \mid \widehat{h}_{n-1}^{(a,b,c)}(i) \mid)\&(\mid \widehat{h}_n^{(a,b,c)}(i) \mid < 10^{-6}); \\ \widehat{h}_n^{(a,b,c)}(i) \ elsewhere. \end{cases} \qquad (12)$$

3.1 Discrete Dual Hahn Moments

The Dual Hahn Moments (DHMs) are scalar quantities formed by using the weighted dual Hahn polynomials. To construct a optimal DHMs for high-resolution images or signal, we invested the method of Daoui et al. explained in the previous section for stable computation of DHPs.

For a signal s(t), with number of samples N, the $n-th$ order of DHMs can be given by the equation below:

$$DHM_n = \sum_{t=0}^{N} \widehat{h}_n^{(a,b,c)}(t)s(t). \qquad (13)$$

The original signal can be obtained by the reverse process:

$$\widehat{s}(t) = \sum_{t=0}^{n} \widehat{h}_n^{(a,b,c)}(t)DHM_n. \qquad (14)$$

For a uniform image $f(x, y)$ with size $N \times N$, the $(n+m)$—th-order dual Hahn moment is defined as follows:

$$DHM_{nm} = \sum_{x=0}^{N}\sum_{y=0}^{N} \widehat{h}_n^{(a,b,c)}(x)\widehat{h}_m^{(a,b,c)}(y)f(x, y). \qquad (15)$$

The reconstructed image $\widehat{f}(x, y)$, at a maximum order possible n_{max} (just the moment up to n_{max} are used), can be approximated by the following relation:

$$\widehat{f}(x, y) = \sum_{x=0}^{n_{max}}\sum_{y=0}^{n_{max}} DHM_{nm}\widehat{h}_n^{(a,b,c)}(x)\widehat{h}_m^{(a,b,c)}(y). \qquad (16)$$

The difference between the reconstructed signal and the original one is measured by the Mean Square Error (MSE) criterion, and the Peak Signal-to-Noise Ratio (PSNR), which are given below:

$$MSE = \frac{1}{N}\sum_{t=1}^{N}(\widehat{s}(t) - s(t))^2. \qquad (17)$$

Knowing that $\widehat{s}(t)$ and $s(t)$, respectively, represent original and reconstructed signal.

$$PSNR = 10 \times log\left(\frac{\sum_{t=1}^{N} s(t)^2}{\sum_{t=1}^{N}(\widehat{s}(t) - s(t))^2}\right) \qquad (18)$$

$$MSE = \frac{1}{N^2}\sum_{i=1}^{N}\sum_{j=1}^{N}(\widehat{f}(i, j) - f(i, j))^2. \qquad (19)$$

Knowing that $\widehat{f}(i, j)$ and $f(i, j)$, respectively, represent the pixel intensity for the original and reconstructed image.

$$PSNR = 10 \times log\left(\frac{\sum_{i=1}^{N}\sum_{j=1}^{N} f(i, j)^2}{\sum_{i=1}^{N}\sum_{j=1}^{N}(\widehat{f}(i, j) - f(i, j))^2}\right). \qquad (20)$$

We summarize the steps of the algorithm for calculating DHMs with respect to the order n in Algorithm 1.

To simplify the complexity and enhance the speed of calculation time of 1D/2D DHMs of order n/(n, m) in MATLAB software, we use the following matrix formulation in Table 2.

Table 2 Matrix formulation for DHMs computing

	Dual Hahn moments	Reconstructed image/signal
Signal	$DHM_n = H_n^T s$	$\widehat{s} = H_n DHM_n$
Image	$DHM_{nm} = H_n^T f H_m$	$\widehat{f} = H_n DHM_{nm} H_m^T$

Algorithm 1 Computation of DHMs

Require: The image/signal and order of moments n;
1: Compute DHPs using equations 11, 10 and 8;
2: Compute DHMs for the input signal/image using the matrix formulation in Table 2;
3: Signal/image reconstruction using the matrix formulation;
4: Measure the mean reconstruction error between the original and reconstructed signal/images ;
Ensure: The reconstructed image/signal using DHMs, MSE, PSNR.

where

$$\begin{aligned} H_n &= \left[\widehat{h}_0^{(a,b,c)}(i), \widehat{h}_1^{(a,b,c)}(i), \ldots, \widehat{h}_{n-1}^{(a,b,c)}(i)\right] \\ H_m &= \left[\widehat{h}_0^{(a,b,c)}(i), \widehat{h}_1^{(a,b,c)}(i), \ldots, \widehat{h}_{m-1}^{(a,b,c)}(i)\right] \end{aligned} \tag{21}$$

denote the matrices of discrete DHPs computed, respectively, up to the orders n and m.

4 Optimized Dual Hahn Moments

This section concerns the proposed method, which is based on firefly optimization algorithm. With automatic and accurate selection of a and c parameters, our method is based on minimizing the mean square reconstruction error (MSE) defined in Eq. (19).

The Dual Hahn Moments (DHMs) are based on parametric DHPs, which suffer from optimal parameter (a, b, and c) selection. As shown in Fig. 1, the parameters affect directly the region of interest in a given set of samples which can be a signal or image. For this reason, we use a swarm optimization technique known as Firefly Algorithm (FA). The mean steps of FA for suitable and optimal image reconstruction by DHMs by tracking the steps below:

Step 1: *Initialization of the Firefly Algorithm (FA) parameters:*
N: Number of fireflies.
X_{max}: Upper bound of decision variables.
X_{min}: Lower bound of decision variables.
It_{max}: Maximum number of iterations.
γ: Absorption coefficient equals 0.01.
β_0: Attractiveness coefficient set to 1.0.
α: Randomization parameter, defined as $\alpha = \alpha_0 \delta$, where $0 < \alpha_0 < 1$ and $0.95 < \delta < 0.97$ for optimal performance.

Step 2: *Define the objective function* $f(X_i)$: For this application, the objective function to minimize is the mean square reconstruction error (MSE) defined before in Eq. (10).

Step 3: *Generate initial firefly positions:* Generate a random set of firefly positions X_i using the continuous uniform distribution, bounded by X_{min} and X_{max}, as

$$X_{ij} = X_{ij} + \Phi(X_{max} - X_{min}), \tag{22}$$

where $\Phi = \text{rand}(0, 1)$ and X_{ij} is the i-th potential solution.

Step 4: *Evaluate light intensity:* Use the objective function $f(X_i)$ to calculate the light intensity for each firefly.

Step 5: *Compare and move fireflies:* For each firefly X_i, compare its light intensity with that of another firefly X_j: If X_j is brighter, move firefly X_i toward X_j using:

$$X_i^{t+1} = X_i^t + \beta_0 e^{-\gamma r_{ij}^2}(X_j^t - X_i^t) + \alpha_t \epsilon_i^t, \tag{23}$$

where ϵ_i^t are random numbers and r_{ij} represents the distance between X_i and X_j.

Step 6: *Update positions:* Update the positions of all fireflies using a greedy selection process to retain the best solutions.

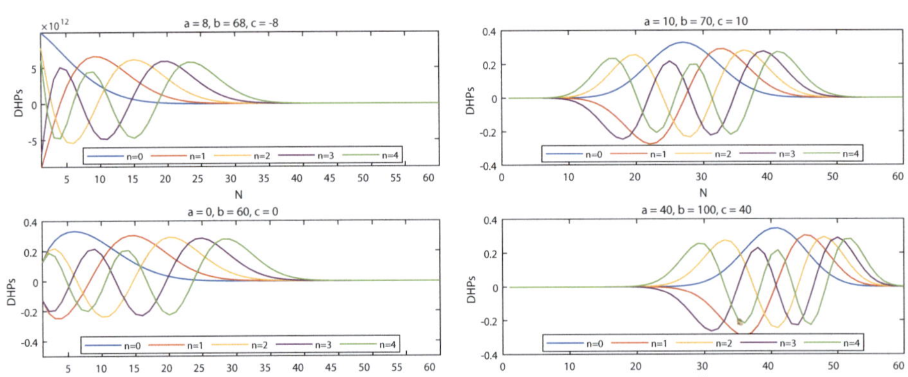

Fig. 1 DHPs as a function of 60 samples from order 0 to 4

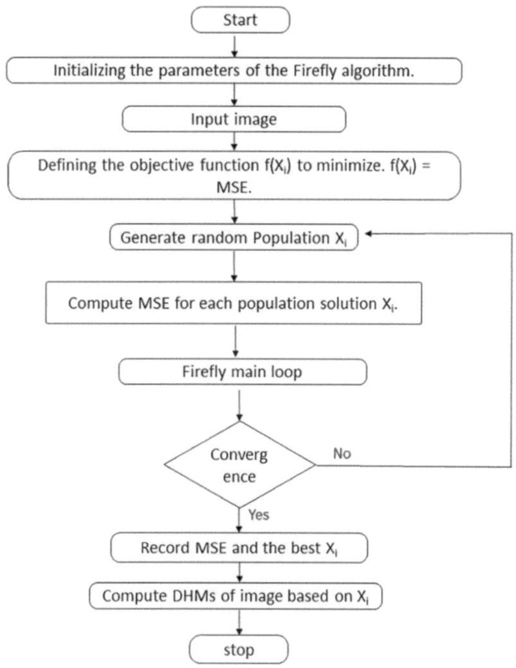

Fig. 2 The flowchart for the proposed method for optimal DHMs

Step 7: *Repeat:* Repeat Steps 3, 4, and 5 until convergence or until the maximum number of iterations ($It_{\max}$) is reached.

Step 8: *Record the optimal solution:* Store the parameter pair $X = (a, b)$ corresponding to the best solution found by the FA.

The parameters in the firefly algorithm were chosen based on their impact on the balance between exploration and exploitation. The absorption coefficient (0.01) was selected to ensure a gradual decay in attractiveness, allowing fireflies to explore a larger search space without premature convergence. The attractiveness coefficient (1) was chosen for simplicity and effectiveness, as it typically provides a good balance in many optimization problems. The randomization parameter (α), with values between 0 and 0.97, introduces controlled randomness, preventing the algorithm from converging too quickly to a local optimum and ensuring sufficient diversity in the solution space. These values were selected based on prior research and preliminary tests, ensuring robust performance for the problem at hand. However, the computational cost of the FA is moderate to slow, but robust in parameter choosing, where it can be not applicable on real-time applications or large-scale datasets. The main steps of the proposed method can be summarized in flowchart which is illustrated in Fig. 2.

5 Experimental Results

This section will evaluate our suggested method for automatically choosing DHM parameters in the first stage, followed by optimal 1D signal and 2D image reconstruction utilizing optimized DHMs in the second stage.

5.1 Automatic Parameter Selection

In this section, we compare two swarm-based optimization algorithms, ABC and FA, for optimizing DHMs in order to achieve optimal image reconstruction. It is important to emphasize that all comparisons are performed under the same conditions: 100 populations, 5 iterations, and 2 dimensions. The curves in Fig. 3 show the MSE and PSNR results comparing the standard method (where parameters are selected manually), the ABC algorithm, and the FA algorithm for the "Castle" image (Fig. 5a). The results indicate that the FA method yields lower reconstruction error (MSE) and higher PSNR, demonstrating its superior performance.

5.2 Optimal 1D Signal and 2D Image Reconstruction

This part presents the performance and efficiency of the proposed method for the optimal and stable reconstruction of large-scale 1D signals and 2D images.

5.2.1 Signal Reconstruction

Figure 4 presents the original one-dimensional signal $s(t)$ with 500 samples, and an amplitude ranging from -400 to 400. Additionally, it shows the reconstructed signal using DHMs for orders $n = 100$, $n = 400$, and $n = 500$, where the parameters a and c are selected using firefly optimization. The parameter b is always chosen as $b = a + N$, where $N = 500$ is the size of the signal $s(t)$.

5.3 Image Reconstruction

In this part the following images in figure are used to test the image reconstruction using optimized DHMs.

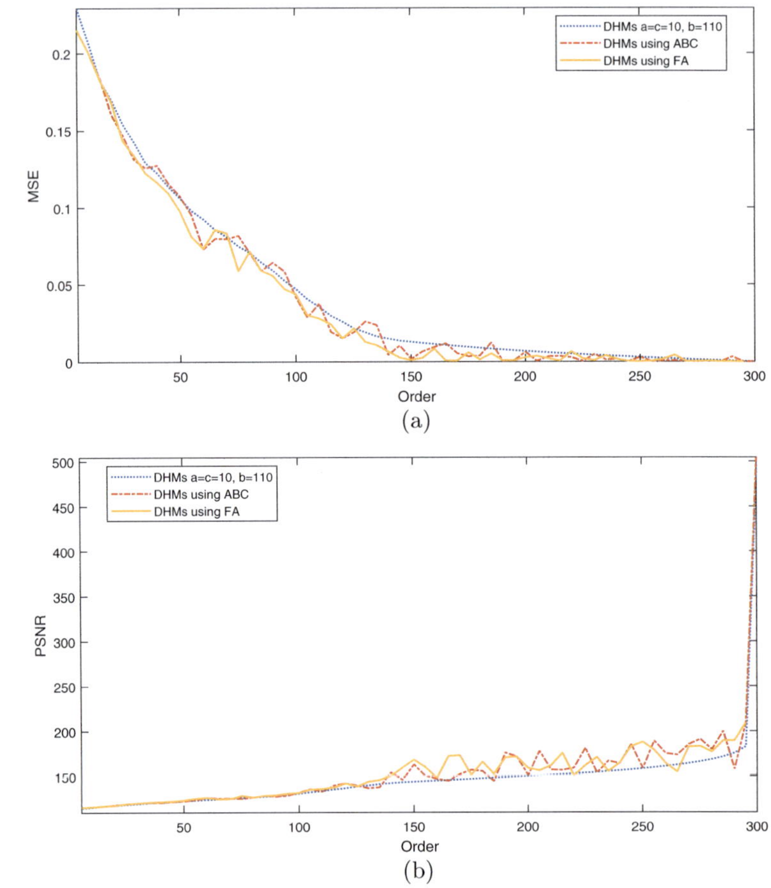

Fig. 3 MSE **a** and PSNR **b** as functions of DHM orders, analyzed across three different parameter selection methods

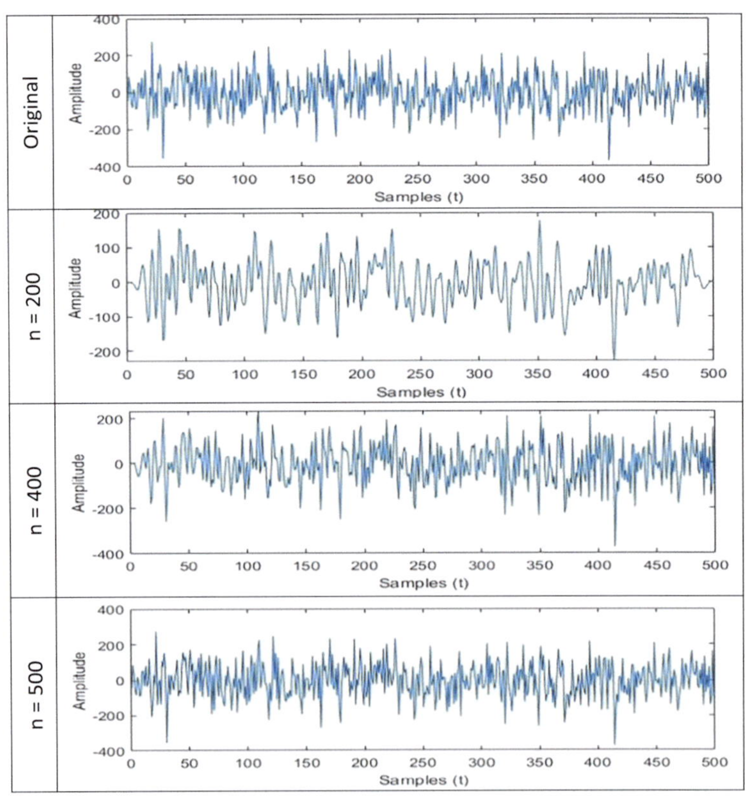

Fig. 4 Reconstruction of signal from DHMs using firefly algorithm for different orders

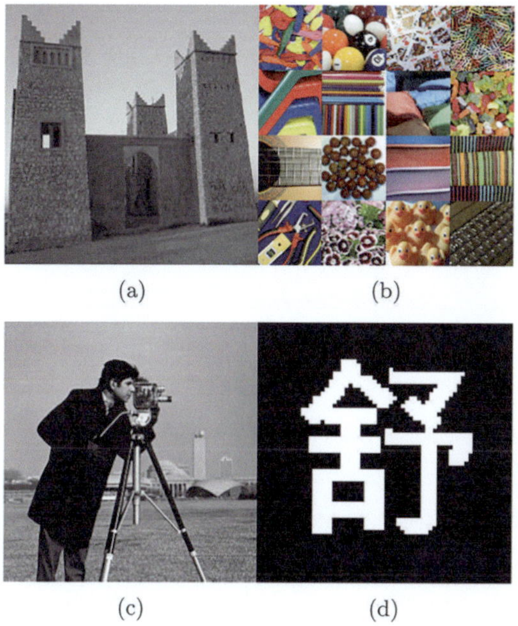

Fig. 5 The test images: **a** "Cameraman" of size 256 × 256. **b**: "Texture" image of size 1024 × 1024. **c**: "MRI" image of size 500 × 500

Figure 6 illustrates the reconstruction of the "Texture" image using both the standard method and the automatic selection method. The best parameters selected by the FA method were $a = 6$, $c = 5$, with $b = 1024 + a$ consistently, and for orders $n = 100$, $n = 200$, and $n = 500$, the selected parameters were $a = 1$, $c = 0$, and $a = 39$, $c = 38$, respectively. Figure 7 presents a comparative analysis of the "Shu" image reconstruction, evaluating the MSE and PSNR between standard DHMs and optimized DHMs. The comparison includes other parametric moments such as Hahn moments (HMs) Zhou et al. [2005], Charlier moments (CMs) Abdul-Hadi et al. [2020], and Meixner moments (MMs) Bencherqui et al. [2022]. The results highlight the superiority of optimized DHMs, particularly at lower orders.

6 Conclusion

In this paper, we propose an algorithm for the automatic selection of Dual Hahn Polynomial (DHP) parameters to construct optimal image and signal moments. This method leverages an optimization approach based on the firefly algorithm, ensuring that the mean square reconstruction error converges to its minimum value. The proposed approach introduces a novel way of selecting DHP parameters, which can vary depending on the size and type of images and signals. Experimental results demonstrate the effectiveness of this method. It is important to note that the adopted optimization algorithm is iterative, meaning it requires some computation time. However, it converges to its global minimum within just a few iterations.

Fig. 6 Texture image reconstruction using FA and standard method for DHPs orders (100, 100), (200, 200), and (500, 500)

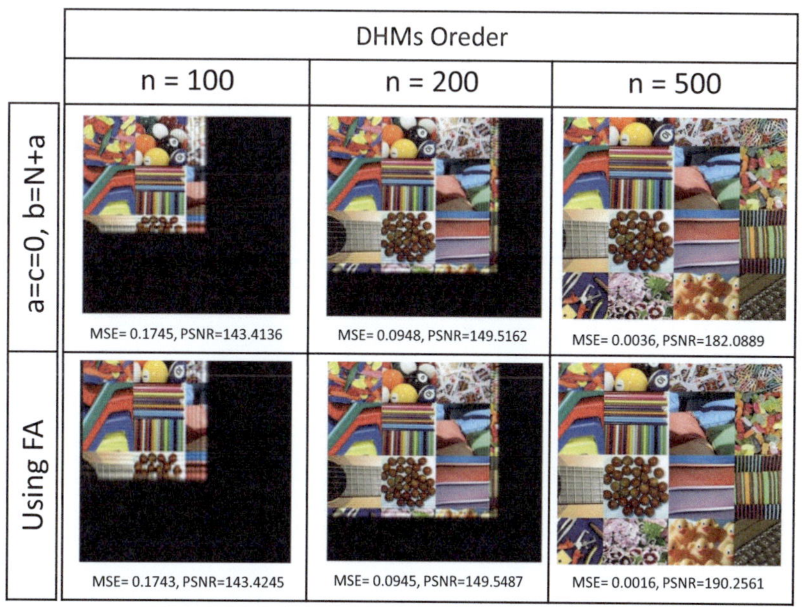

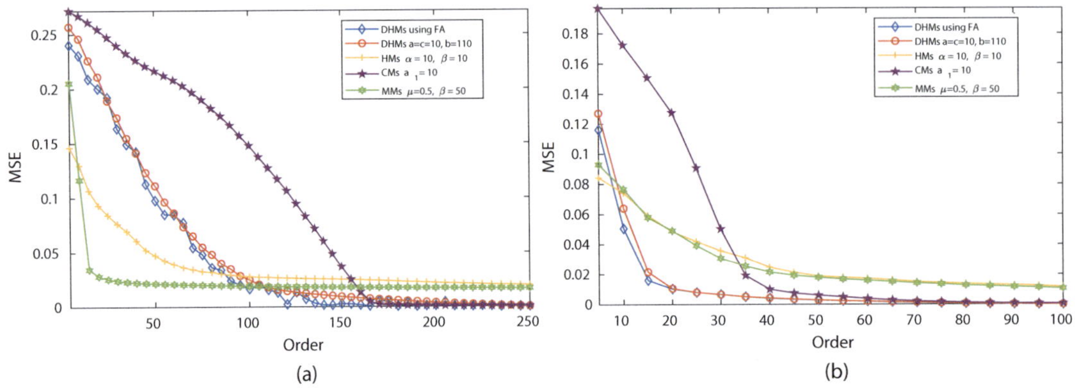

Fig. 7 MSE as a function of different orthogonal moments order for: **a** Grayscale "Cameraman" image. **b**: Binary "Shu" image

References

Abdul-Hadi, A.M., Abdulhussain, S.H., Mahmmod, B.M.: On the computational aspects of charlier polynomials. Cogent Engineering **7**(1), 1763553 (2020)

Asli, B.H.S., Flusser, J.: Fast computation of krawtchouk moments. Information Sciences **288**, 73–86 (2014)

Bencherqui, A., Tahiri, M.A., Karmouni, H., Daoui, A., Alfidi, M., Jamil, M.O., Qjidaa, H., Sayyouri, M.: Optimization of meixner moments by the firefly algorithm for image analysis. In: International Conference on Digital Technologies and Applications, pp. 439–448 (2022). Springer

Bourzik, A., Bouikhalene, B., El-Mekkaoui, J., Hjouji, A.: A comparative study and performance evaluation of discrete tchebichef moments for image analysis. In: Proceedings of the 6th International Conference on Networking, Intelligent Systems & Security, pp. 1–7 (2023)

Bourzik, A., Bouikhalene, B., El-Mekkaoui, J., Hjouji, A.: Accurate image reconstruction by separable krawtchouk-charlier moments with automatic parameter selection using artificial bee colony optimization. Multimedia Tools and Applications, 1–22 (2024)

Bourzik, A., Bouikhalene, B., El-Mekkaoui, J., Hjouji, A.: Advanced optimization of discrete orthogonal hahn moments for precision image reconstruction via the firefly algorithm. In: 2024 3rd International Conference on Embedded Systems and Artificial Intelligence (ESAI), pp. 1–9 (2024). IEEE

Bourzik, A., Bouikhalene, B., El-Mekkaoui, J., Hjouji, A.: Invariant tchebichef and krawtchouk moments for shape recognition: A performance evaluation study. In: Proceedings of the 7th International Conference on Networking, Intelligent Systems and Security, pp. 1–8 (2024)

Daoui, A., Karmouni, H., Yamni, M., Sayyouri, M., Qjidaa, H.: On computational aspects of high-order dual hahn moments. Pattern Recognition **127**, 108596 (2022)

Dey, N., Chaki, J., Moraru, L., Fong, S., Yang, X.-S.: Firefly algorithm and its variants in digital image processing: A comprehensive review. Applications of firefly algorithm and its variants: case studies and new developments, 1–28 (2020)

El Mallahi, M., Zouhri, A., Mesbah, A., Qjidaa, H.: 3d radial invariant of dual hahn moments. Neural Computing and Applications **30**, 2283–2294 (2018)

Hjouji, A., Bouikhalene, B., EL-Mekkaoui, J., Qjidaa, H.: New set of adapted gegenbauer–chebyshev invariant moments for image recognition and classification. The Journal of Supercomputing **77**, 5637–5667 (2021)

Mesbah, A., Zouhri, A., El Mallahi, M., Zenkouar, K., Qjidaa, H.: Robust reconstruction and generalized dual hahn moments invariants extraction for 3d images. 3D Research **8**, 1–20 (2017)

Papakostas, G.A., Karakasis, E.G., Koulouriotis, D.E.: Efficient and accurate computation of geometric moments on gray-scale images. Pattern Recognition **41**(6), 1895–1904 (2008)

Raj, P.A.: Charlier and meixner moments and their application for texture and image de-noising problems. In: Information, Photonics and Communication: Proceedings of Second National Conference, IPC 2019, pp. 123–134 (2020). Springer

Sharma, A., Sehgal, S.: Image segmentation using firefly algorithm. In: 2016 International Conference on Information Technology (InCITe)-the Next Generation IT Summit on the Theme-internet of Things: Connect Your Worlds, pp. 99–102 (2016). IEEE

Yahya, S., Jaouad, E.-M., BENSLIMANE, M., Idrissi, B.J., El Ogri, O., *et al*: Optimal reconstruction and recognition of images by jacobi fourier moments and artificial bee colony (abc) algorithm. Statistics, Optimization & Information Computing **12**(3), 829–840 (2024)

Zhou, J., Shu, H., Zhu, H., Toumoulin, C., Luo, L.: Image analysis by discrete orthogonal hahn moments. In: International Conference Image Analysis and Recognition, pp. 524–531 (2005). Springer

Zhu, H., Shu, H., Zhou, J., Luo, L., Coatrieux, J.-L.: Image analysis by discrete orthogonal dual hahn moments. Pattern Recognition Letters **28**(13), 1688–1704 (2007)

Zhu, H., Shu, H., Liang, J., Luo, L., Coatrieux, J.-L.: Image analysis by discrete orthogonal racah moments. Signal Processing **87**(4), 687–708 (2007)

A Comparative Study of Decentralized and Distributed Architectures for Deploying Federated Learning Models

Saad Mahmoudi and Mohamed Amnai

Abstract

The continuous generation of massive amounts of data from connected smart objects, autonomous systems, and embedded systems, in real or delayed time, requires significant storage resources, as well as powerful computing resources and high-performance control and security systems. In this article, we will discuss the various research projects conducted by researchers with the aim of developing distributed, decentralized, and secure federated learning architectures. We begin with cloud-based FL architecture and an edge-based FL architecture, both of which allow IoT devices to leverage storage and computing resources for the reliable and efficient processing of decentralized learning models. Next, we introduce federated learning (FL) in an AI-based cloud architecture, which involves the use of an intelligent data transfer system to improve computation time and reduce latency between the cloud server and IoT devices. Then we introduce a modularized federated learning architecture for privacy and security considerations, which aims to ensure data security and confidentiality and create a privacy-enhancing process for data exchange between IoT devices. Finally, we introduce an architecture that integrates the learning model into a blockchain system while applying a POW (proof of work) consensus mechanism that guarantees the integrity and confidentiality of the hyperparameters of local learning models. Despite all the critical issues addressed by deploying FDL in these above-mentioned architectures, decentralized and federated learning still faces challenges such as improving the training time of learning models, security of data and learning models, data heterogeneity issue, and embedded resources on edge devices. This leads us to design a new secure and distributed federated learning architecture based on new lightweight and secure network and telecommunication technologies.

Keywords

Federated learning · Federated edge learning · Decentralized architecture distributed learning · Blockchain · MQTT · Broker MQTT

1 Introduction

Considering the technological evolution of computing and storage chips embedded in various smart devices, as well as the increasing amount of data continuously generated by these systems, IoT infrastructure requires very high storage capacities, energy capacities, and security systems during the training phase of a learning model. Distributed model learning relies on the dataset, which must be maintained on participatory nodes during the training phase of the learning model (Liuetal 2020). The learning results are then sent to the cloud server. Therefore, managing the distribution of learning parameters based on the non-independent identical distribution (non-IID) situation is mandatory in the infrastructure of a smart city ecosystem (Zhao et al. 2018).

Although data and training parameter processing times are improving thanks to computing resources, storage, and parameter protection services, these performances allow learning models to achieve a high level of intelligence (Kehoeetal 2015). Other challenges to overcome include the heterogeneity of IoT infrastructure, data security, and network congestion due to the exchange of learning parameters between different connected devices. For these reasons,

S. Mahmoudi (✉) · M. Amnai
Ibn Tofail University of Kenitra, Kenitra, Morocco
e-mail: saad.mahmoudi@uit.ac.ma

M. Amnai
e-mail: mohamed.amnai@uit.ac.ma

M. Gouskiret al. (eds.), *Cutting-Edge Artificial Intelligence Applications*, Sustainable Artificial Intelligence-Powered Applications,
https://doi.org/10.1007/978-3-032-00125-2_5

we considered adopting a federated, shared, and secure learning approach to monitor any flow of learning parameters between connected objects and accelerating the shared learning of learning models. Connected objects always present the risk of constantly sharing and exchanging data between themselves due to the unreliability and insecurity of the network infrastructure dedicated to data exchange between devices and aggregation servers. Some researchers have considered preserving the confidentiality of sensitive data from connected objects so that they are neither lost nor tampered with by hackers (Zhu et al. 2019).

In this article, we will dissect the five architectures that cover an aspect of training security and learning model processing speed, each of which attempts to address security, resilience, and latency issues during the training phase of a federated learning model.

2 Related Work

Several research projects are being conducted for the optimal deployment of federated learning models in various distributed and secure architectures. These projects aim to accelerate the learning model training phase, minimize data loss, ensure the integrity of the learning parameters of connected objects, and minimize the risk of hardware resource failure of a connected object, as well as network traffic overload. The exponential emergence of new autonomous objects and systems in the smart city environment shows that the scalability of distributed architecture is one of the key approaches for developing a learning model that is more resilient to changes in the state of objects and even the environments to which they belong, taking into account the following critical elements: balancing the data processing load, the interoperability of connected objects, and their heterogeneity.

In this section, we briefly present the different avenues of research carried out for the evolution and development of distributed, decentralized, and secure machine learning architectures.

2.1 Cloud and Edge-Cloud-Based Architecture of Federated Learning

The federated learning deployment architectures as shown in Fig.1 and Fig.2, allows client devices to take advantage of the hardware infrastructure in terms of computing and processing power in the training phase of learning model such as CPU, GPU, and TPU because the efficiency of resource allocation (RA) is possibly influenced by the implementation of the federated learning (FL) system, especially the bandwidth which is essential when exchanging learning parameters, and by the use of advanced technologies, including software-defined networks (SDN) and virtual network functions (VNF) by cloud providers for the management of computing and communication resources, these problems of limited resources at the IoT device level and network congestion have been solved.

First, the fundamental concept of federated learning is to reduce the communication load by sending only the model parameters instead of large raw datasets (Herabad 2023). Additionally, leveraging client processing capabilities and minimizing data transport distance helps optimize resource utilization and reduce latency (Rahman et al. 2023). Wang et al. (2021) addressed the aspect of bandwidth management, which is a key and critical element during the model training phase of the federated learning system, by making improvements to the allocation process by predicting bandwidth requirements and selecting the optimal path. For efficient storage management, Liu demonstrated in Liu et al. (2022) that federated learning (FL) requires much less storage space than traditional machine learning because the raw data is collected and processed on the client's device, not on the storage server, maintaining the confidentiality of data and training parameters. However, the current research landscape on cloud-based and edge computing federated learning (FL) architectures demonstrates that there are still challenges focused on security, scalability, data integrity, and trust establishment.

The bandwidth overhead resulting from the constant exchange of training data between client devices and the cloud can strain network resources and lead to increased energy consumption. To reduce communication delays and implementation costs in the cloud, Wang et al. proposed in Wang et al. (2021b) a unitary over the air computation (UMAirComp) framework to facilitate efficient edge federated learning. However, data exchange over a network infrastructure such as a Wide Area Network (WAN) still poses problems in terms of integrity and confidentiality. To overcome these challenges, other alternatives exist, such as the work of Wu et al. (2020), where custom models are created on clients and global models on aggregation servers at the cloud level. Another alternative to exploiting cloud-edge capabilities is hierarchical architecture, in addition to client clustering for scalability (Abdellatif et al. 2022).

Both cloud and edge FL architectures share the same drawbacks: a single point of failure and high communication costs. Therefore, research focuses on deploying federated learning systems in architectures based on distributed methods among the clients themselves (Lalitha et al. 2018; Attanayaka et al. 2023) and (Pappas et al. 2021). FL architectures that distribute computation without blockchain technology still face challenges such as communication overhead for model updates and the security of data and learning

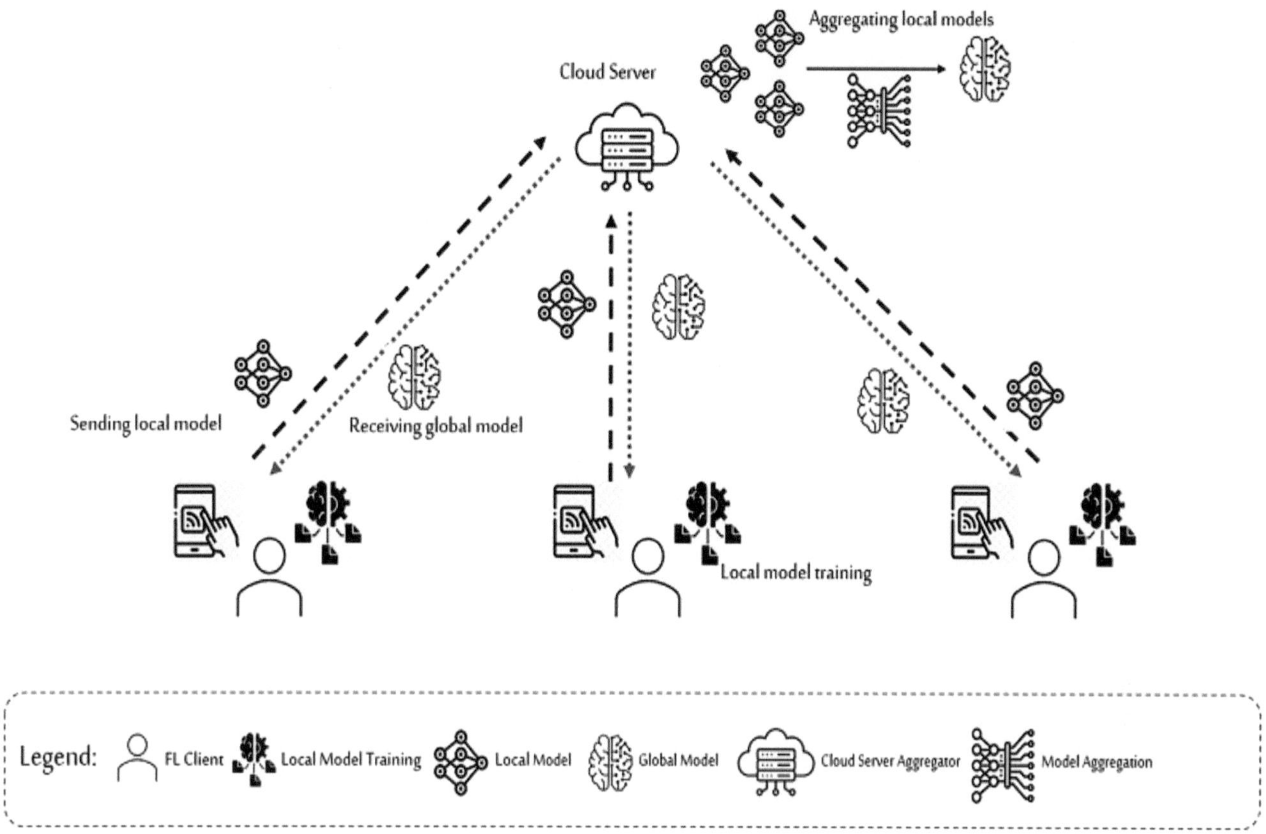

Fig. 1 Cloud-based FL architecture (Alnajar 2024)

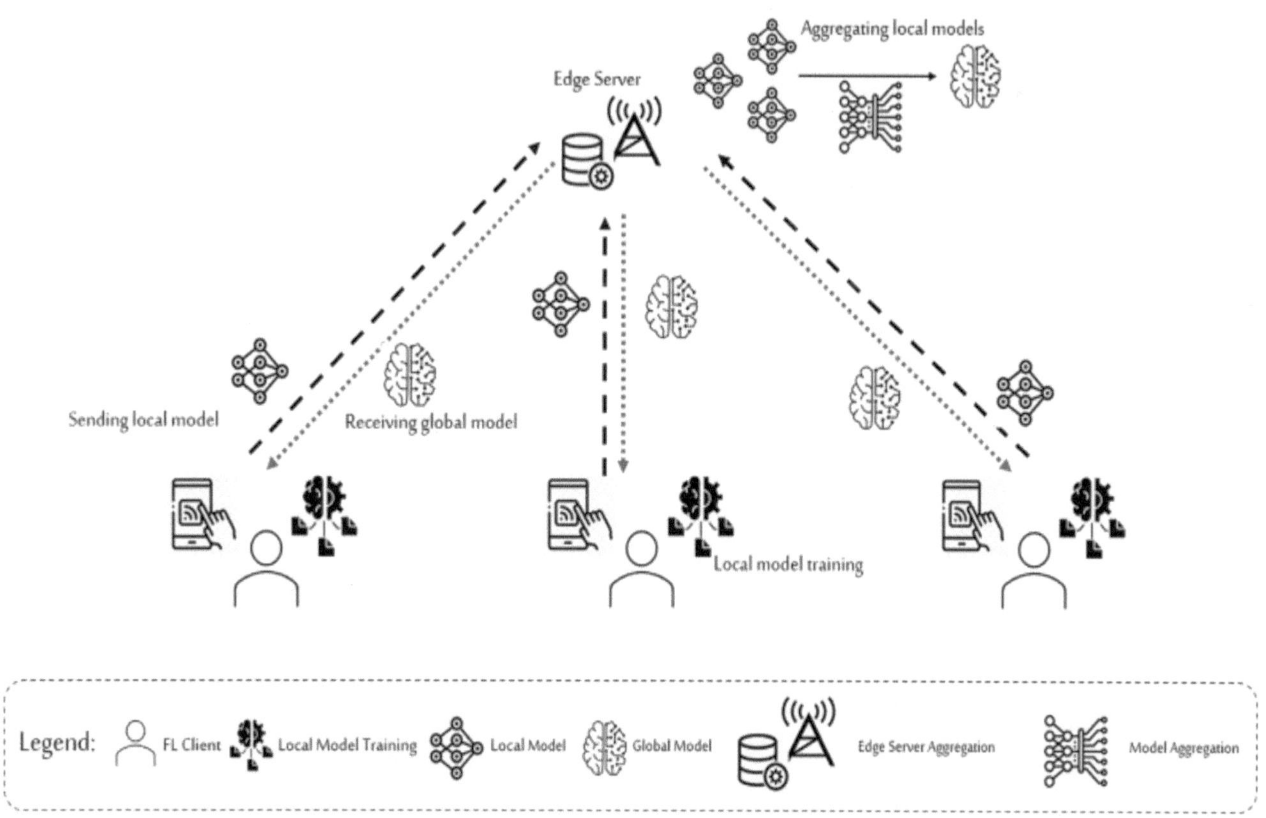

Fig. 2 Edge-based FL architecture (Alnajar 2024)

parameters exchanged between IoT clients and the aggregation server (Figs. 1 and 2).

2.2 Federated Learning in Edge-Cloud Architecture Based on AI

The architecture, as depicted in Fig. 3, consists of performing local and global training of federated learning models, by creating an intelligent transfer system of learning models between IoT devices and cloud servers through a network layer. It plays a key role in reducing latency and reliable data processing and also in improving quality of service. This architecture includes four layers: the edge layer, the network layer, the cloud layer, and the application layer which are described as follows:

Edge Layer: This is the fundamental layer which includes the edge nodes and edge controllers. In this layer, a large volume of data generated or collected by IoT devices from their environment being subjected to training via the local federated learning model. This local model is then transmitted to the peripheral controller, where it is merged with other local models to generate the intermediate model. The edge layer is essential to reduce the processing time as well as improve the quality of service of AIoT devices (artificial intelligence into IoT devices) (AbdulRaheem et al. 2022).

Network layer: It plays a vital role in ensuring secure and reliable wired and wireless transmission of the generated models between the device controller and cloud servers by avoiding network issues such as congestion via IP, OSPF, and BGP routing protocols and contributes to the overall quality of service of the proposed system (Gouveia and Magedanz 2009).

Cloud Layer: Within this layer, cloud servers aggregate the training model sent by device controllers to generate a global model that is then sent back to the devices. The cloud layer plays a vital role in optimizing the quality of service of AIoT devices (artificial intelligence into IoT devices) by exploiting the scalability and computing power of cloud computing (Koubâa et al. 2020).

Application layer: The upper layer is the interactive layer with end users, which is responsible for offering services such as a navigation system that provides users with precise information that can help them make adequate decisions.

This architecture is designed to optimize the quality of service of AIoT devices (artificial intelligence into IoT devices) taking advantage of the services that can be provided by the edge-cloud and applied in various AIoT (artificial intelligence into IoT devices) applications, including intelligent transportation systems (Rong et al. 2021) (Fig. 3).

2.3 A Federated Learning Architecture Modularized for Privacy and Security Considerations

This architecture, as depicted in Fig. 4,defines and decomposes the modularized ML process into independent units and keeps the data on local client devices providing an additional layer of confidentiality to the clients. This layer helps preserve and protect not only the data of the connected devices but also the parameters of their learning models. It also serves to ensure the integrity of the learning parameters so that they are neither compromised nor falsified by attacks that can also bias the global model. These attacks can also generate inferences that favor a particular entity or party (Banabilaha et al. 2022) (Fig. 4).

In the same context, authors Xu, Li, Liu, Yang, and Lin proposed in Banabilaha et al. (2022) a process called VerifyNet to solve two problems of deep learning networks. The first problem was to preserve the confidentiality of user information during the training process, and the second was to verify the correctness of the model results' (or predictions) broadcast by the server. For this, the aggregation server needs to calculate the aggregated results along with the corresponding proof for each user. If the verification is successful, the distribution of the uploaded data used to generate the aggregated results must be the same as the real data. This also confirms that a result returned by the server is correct once its proof is verified (Banabilaha et al. 2022; Guowen et al. , 2019).

2.4 Architecture of a Federated Learning System Based on Hierarchical Blockchain

In this section, we present the hierarchical blockchain for federated learning on non-IID datawhere a blockchain-based system that allows data to be distributed across different connected objects. It also ensures the sharing of learning models aggregated by the server between these devices. Regarding the blockchain, it presents the basis of this architecture which has the role of securing, encrypting, and validating transactions written in data blocks by miners (Zhang et al. 2024). It is also used in a process that serves to facilitate the aggregation of global models and to adapt it to all users (nodes, autonomous systems, and connected objects) as shown in the architecture of the federated learning system based on the hierarchical blockchain in Fig. 5.

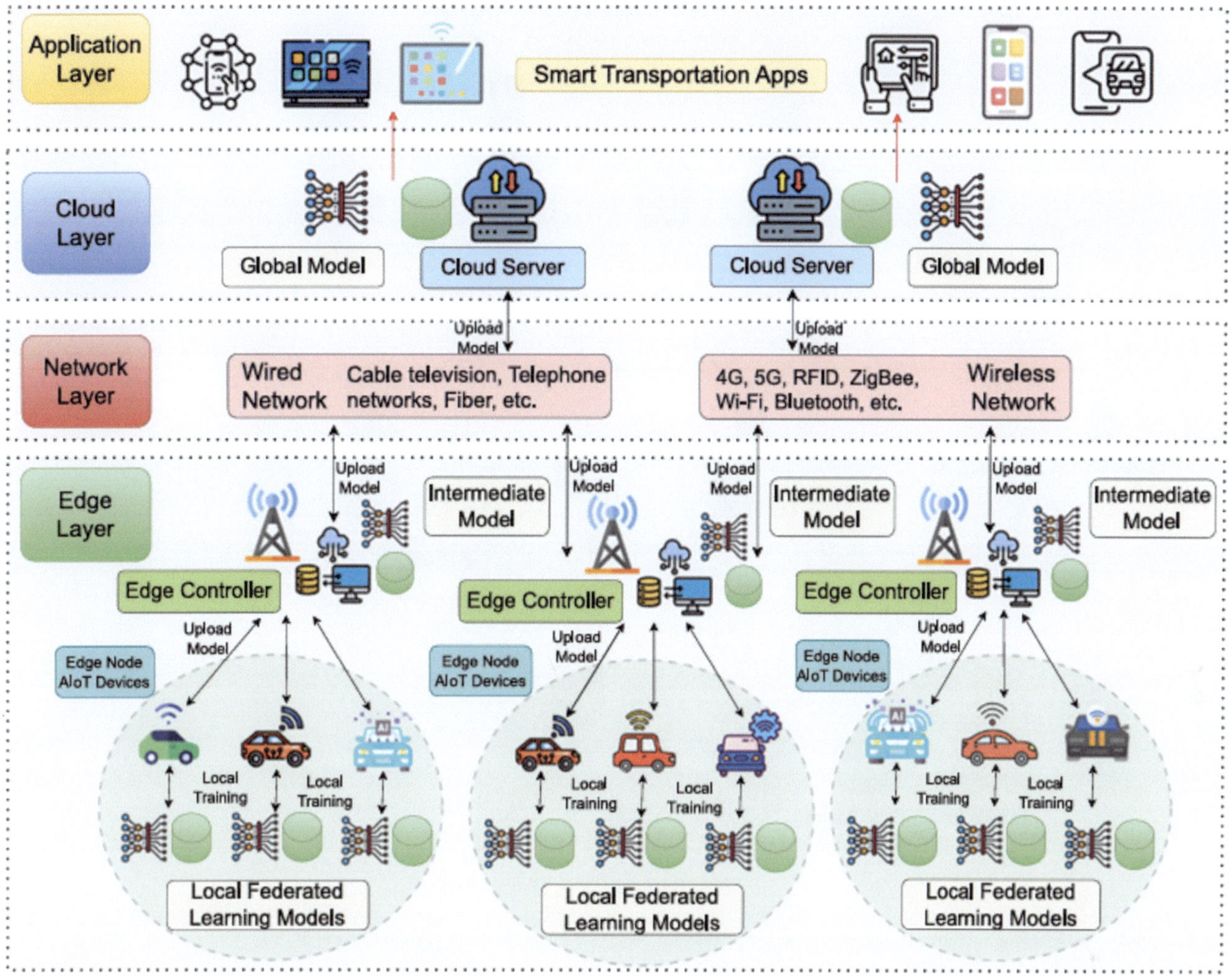

Fig. 3 Federated learning in edge-cloud architecture based on AI (Zhou et al. 2024)

The architecture illustrated in Fig. 5 is composed of cloud server, main blockchain, blockchain system, and smart contract which are as follows (Zhang et al. 2024).

Cloud server allows to aggregate the parameters of the local learning models of the different nodes (connected objects) after a cross validation ensured by the blockchain miners to subsequently generate a global model that will be distributed to the nodes located in the federated learning architecture.

Main blockchain is the decentralized network that is responsible for applying a POW (proof of work) consensus algorithm to validate the writing of the parameters of the learning models that are sent by the users (nodes).

Smart contract is an automated computer program built on blockchain technology. This can automatically execute tasks and control various operations and behaviors specified in the terms of the contract. Smart contracts often provide incentives for task completion while relying on decentralization, automation, and trust mechanisms, and play an important role in multi-party collaboration and transactions. The proposed use of smart contracts in this paper is based on the Block FL architecture presented in Kim et al. (2019) for horizontal federated learning.

The operational steps of the smart contract proposed in Zhang et al. (2024) are as follows:

Step 1: The authority sets the current federated learning cycle e and the initial model M0. The smart contract sends the initial model M0 to clients, who train the initial model M0 with their local data to obtain the local model Mn.

Step 2: Each client uploads the Mn to the smart contract, which performs a proof of workload for the model accuracy based on its own test set and issues an on-chain permit. If the time of a federated learning cycle has elapsed or enough local models have been collected, the smart contract sends an aggregation signal to the server.

Step 3: The server retrieves the queue of local models for a federated learning cycle from the blockchain, applies the federated averaging algorithm to aggregate the local

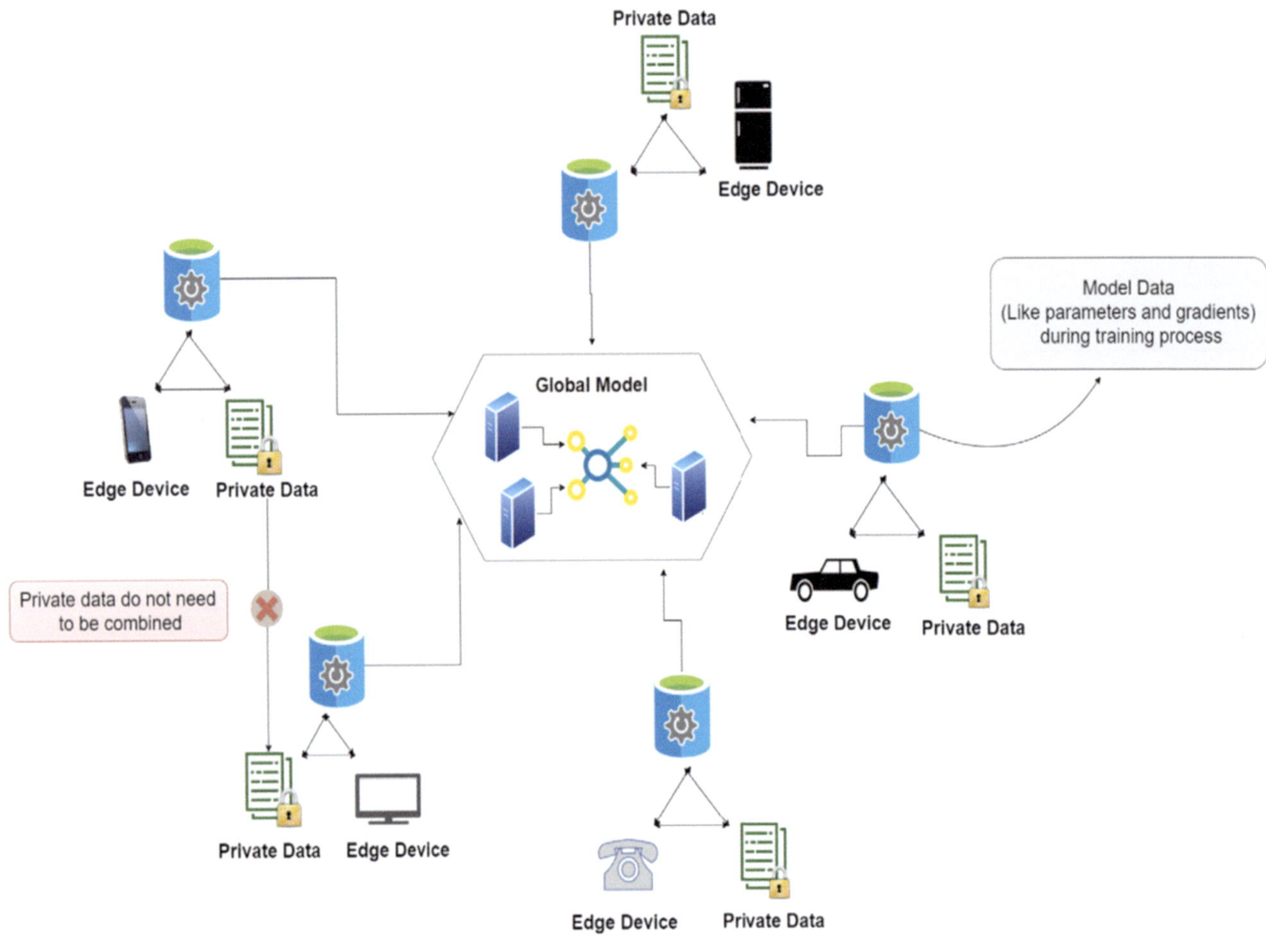

Fig. 4 A federated learning architecture modularized for privacy and security considerations (Banabilaha et al. 2022)

Fig. 5 The architecture of a federated learning system based on hierarchical blockchain (Zhang et al. 2024)

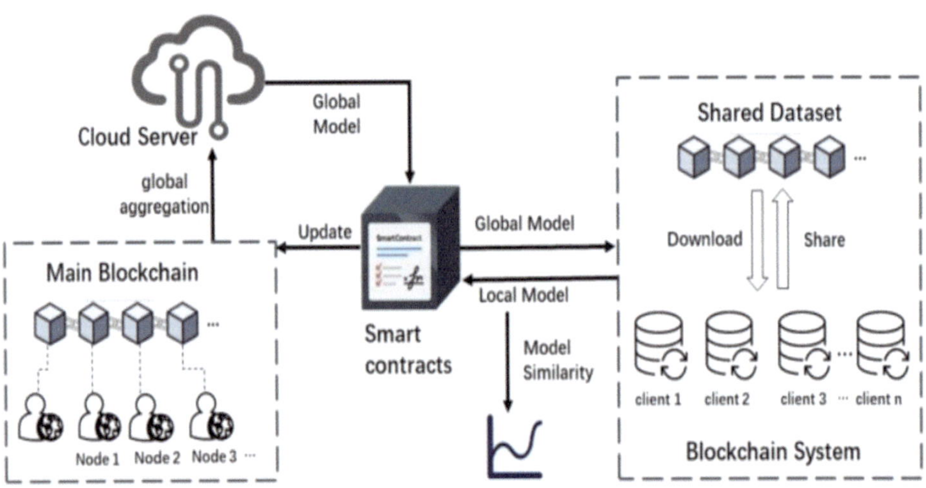

models, obtains a global model, and then stores it in the main blockchain.

Step 4: The smart contract sends a signal to the clients to start the next round of decentralized federated learning using the aggregated global model.

3 Results and Discussions

All experiments were conducted on a laptop equipped with an Intel® Core™ i5-8250U processor clocked at 1.60 GHz (8 cores) and an NVIDIA GeForce 1650 Ti 4 GB graphics

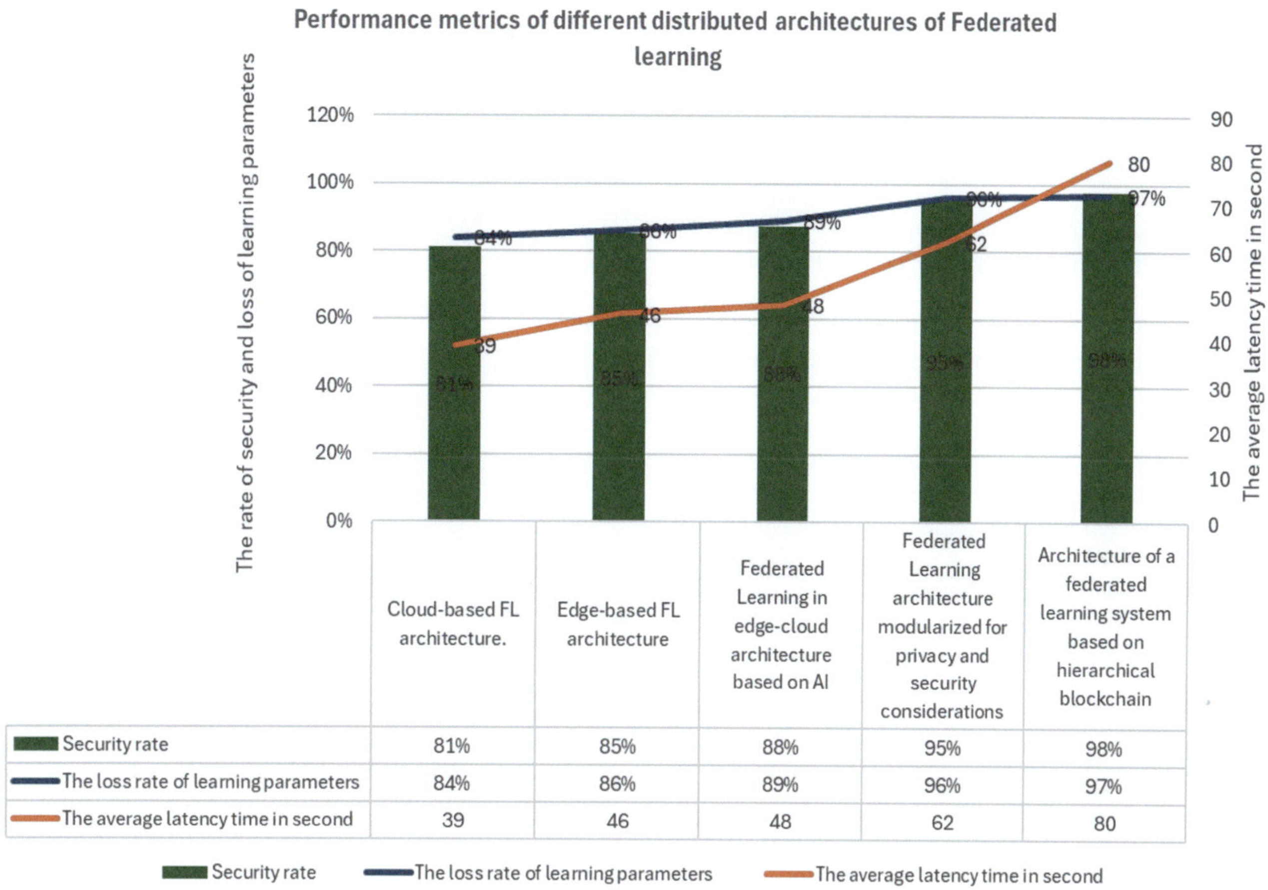

Fig. 6 Graphical representation of performance metrics of different distributed architectures of federated learning

card, using TensorFlow 2.4.0, CUDA 11.0.3, and Keras 2.4.0 from Google Cloud. These resources allow us to simulate the deployment of the federated learning model in the various architectures mentioned above. We also used blockchain as a cloud service to implement federated learning in the decentralized blockchain architecture.

In this comparative study, we used the Flower framework, dedicated to the development of FL systems, as it provides a stable framework for the independent synthesis of key components and high-level interfaces of FL systems. Flower allows FL implementations to be extended to remote and virtual clients with various computing, storage, and network capabilities (Ntantiso et al. 2022).

According to the analysis of the processing results of the learning model training phase with the different architectures that are mentioned in this article, we can deduce their advantages and disadvantages. Regarding the deployment of federated learning in AI-based edge computing, it has an advantage that all IoT objects benefit from very powerful computing resources, and even storage resources, via a very robust infrastructure including a dynamic network layer. Indeed, smart objects receive the initial model from the cloud server, each of them performs local training and sends their learning parameters to the server for aggregation. This means that the generated model may be uncertain or even non-confidential, which poses a problem of efficiency of the training phase of the model based on federated learning. Regarding the modularized federated learning architecture, it guarantees the confidentiality of connected devices and the integrity of the data, learning model parameters. But this process can generate a latency problem during the conformity verification phase of the models exchanged between client IoT devices, and even the aggregated model generated by the server. Regarding the latest federated learning architecture based on a hierarchical blockchain system, the objective is to secure the exchange of learning parameters between connected objects thanks to a POW (proof of work) consensus algorithm. It also uses smart contracts that allow the automation of all tasks and learning processes related to decentralization and confidentiality mechanisms. However, these processes generate the increase in the processing time of learning models, the validation time of parameters, and latency. Therefore, we present the performance indicators of each architecture mentioned in this article in the form of graphs, as shown in Fig. 6.

From the results obtained during the analysis of the different federated learning model architectures mentioned in this article, it is observed that each of them presents performance improvements in terms of security and confidentiality, particularly in terms of learning parameters. Indeed, it is observed that there are performance degradations in terms of latency and processing time granted to connected objects to better optimize the learning cycle of the federated learning model. On the other hand, it is clearly seen that a modular federated learning architecture and a hierarchical federated learning model architecture based on blockchain prove to be advantageous in terms of security during the learning phase of the FL model, but with an increase in latency, which degrades the performance of the federated learning phase. Regarding the three federated learning architectures that are generally based on cloud infrastructure (cloud, edge-cloud or AI-based edge-cloud) as shown in the graph above, each of them has an advantage in terms of latency reduction that accelerates the processing of learning models from connected devices with a fairly low security rate, which causes significant data loss during the training phase.

4 Conclusion and Future Work

In this article, we have presented the different architectures covering the aspects of confidentiality and data integrity, as well as the learning models of connected objects. This comparative study reveals that there are still problems that degrade the performance of the learning model training phase. Therefore, we will propose in future work a new federated, distributed, and clustered learning architecture using an MQTT and HTTPS communication protocol, in order to optimize the learning model training phase. It will also strengthen security, reduce latency, and intelligently reduce network traffic overload during the training phase.

References

Abdellatif , A.A., Mhaisen, N., Mohamed, A., Erbad, A., Guizani, M., Dawy, Z., Nasreddine, W., 2022. Communication-efficient hierarchical federated learning for IoT heterogeneous systems with imbalanced data. Future Gener. Comput. Syst. 128, 406–419.

M. AbdulRaheem, I.D. Oladipo, A. González-Briones, J.B. Awotunde, A.R. Tomori, R.G. Jimoh, An efficient lightweight speck technique for edge-IoT-based smart healthcare systems, in: 5G IoT and Edge Computing for Smart Healthcare, Elsevier, 2022, pp. 139–162.

Omar Alnajar , Ahmed Barnawi Faculty of Computing and Information Technology, King AbdulAziz University, Jeddah, Saudi Arabia, Engineering Applications of Artificial Intelligence 136 (2024) 108934.

Attanayaka, D., Porambage, P., Liyanage, M., Ylianttila, M., 2023. Peer-to-peer federated learning based anomaly detection for open radio access networks. IEEE.

Syreen Banabilaha, Moayad Aloqailyb, Eitaa Alsayeda, Nida Malika, Yaser Jararweha ,Federated learning review: Fundamentals,enabling technologies, and future applications, Information Processing & Management, Elsevier ,2022, 103061.

F.C. de Gouveia, T. Magedanz, Quality of service in telecommunication networks, Telecommun. Syst. Technol. 2 (5) (2009) 77–84.

Guowen Xu, Student Member, IEEE, Hongwei Li (Corresponding author), Senior Member, IEEE, Sen Liu, Student Member, IEEE, Kan Yang, Member, IEEE, Xiaodong Lin, Fellow, IEEE, VerifyNet: Secure and Verifiable Federated Learning, IEEE Transactions on Information Forensics and Security,2019.

Herabad, M.G., 2023. Communication-efficient semi-synchronous hierarchical federated learning with balanced training in heterogeneous IoT edge environments. Internet Things 21, 100642.

B.Kehoeetal.,A survey of research on cloud robotics and automation, IEEE Transactions on automation science and engineering (2015).

Kim H. Park J. Bennis M. Kim S.-L. Blockchained on-device federated learning. IEEE Commun. Lett. 2019; 24: 1279-1283.

A. Koubâa, A. Ammar, M. Alahdab, A. Kanhouch, A.T. Azar, Deepbrain: Experimental evaluation of cloud-based computation offloading and edge computing in the internet-of-drones for deep learning applications, Sensors 20 (18) (2020) 5240.

Lalitha , A., Shekhar, S., Javidi, T., Koushanfar, F., 2018. Fully decentralized federated learning. In: Third Workshop on Bayesian Deep Learning. NeurIPS, Vol.2.

Liu , J., Huang, J., Zhou, Y., Li, X., Ji, S., Xiong, H., Dou, D., 2022. From distributed machine learning to federated learning: a survey. Knowldge Inform. Syst. 64 (4), 885–917.

Y.Liuetal.,Federated learning for 6g communications : Challenges, methods, and future directions, China Communications (2020).

L Ntantiso, A Bagula, O Ajayi, F Kahenga-Ngongo , A review of federated learning: algorithms, frameworks and applications (2022 Dec 5), pp. 341–57.

Pappas, C , Chatzopoulos, D., Lalis, S., Vavalis, M., 2021. Ipls: A framework for decentralized federated learning. In: 2021 IFIP Networking Conference. IFIP Networking, IEEE.

Rahman , A., Hasan, K., Kundu, D., Islam, M.J., Debnath, T., Band, S.S., Kumar, N.,2023. On the ICN-IoT with federated learning integration of communication: Concepts, security-privacy issues, applications, and future perspectives. Future Gener. Comput. Syst. 138, 61–88.

G. Rong, Y. Xu, X. Tong, H. Fan, An edge-cloud collaborative computing platform for building AIoT applications efficiently, J. Cloud Comput. 10 (2021) 1–14.

Wang , X., Hu, J., Lin, H., Garg, S., Kaddoum, G., Piran, M.J., Hossain, M.S., 2021c. QoS and privacy-aware routing for 5G-enabled industrial internet of things: A federated reinforcement learning approach. IEEE Trans. Ind. Inform. 18 (6), 4189–4197.

Wang , S., Hong, Y., Wang, R., Hao, Q., Wu, Y.-C., Ng, D.W.K., 2021b. Edge federated learning via unit-modulus over-the-air computation (extended version). arXiv preprint arXiv:2101.12051.

Wu , Q., Chen, X., Zhou, Z., Zhang, J., 2020. Fedhome: Cloud-edge based personalized federated learning for in-home health monitoring. IEEE Trans. Mob. Comput. 21 (8), 2818–2832.

Feng Zhang. Yongjing Zhang. Shan Ji. Zhaoyang Han. Secure and decentralized federated learning framework with non-IID data based on blockchain. February 29, 2024

Yue Zhao, Meng Li, Liangzhen Lai, Naveen Suda, Damon Civin, Vikas Chandra, Federated learning with non-iid data, 2018, arXiv preprint arXiv:1806.00582.

Jingwen Zhou [a], Shantanu Pal [a], Chengzu Dong [a], Kaibin Wang, , Enhancing quality of service through federated learning in edge-cloud architecture, Ad Hoc Networks, Volume 156, 1 April 2024, 103430.

Ligeng Zhu, Zhijian Liu, Song Han, Deep leakage from gradients, Adv. Neural Inf. Process. Syst. 32 (2019).

Consensual FMEA in Interval-Valued Intuitionistic Fuzzy Environment

Mohamed El Alaoui

Abstract

Failure Modes and Effects Analysis (FMEA) is a method that has long proven its worth in various fields. It consists of evaluating and prioritizing risks to better control their potential impacts. Since no expert can proclaim a perfect knowledge of all the systems assessed, the evaluation process considers several experts' opinions. Hence, how a consensus can be drawn? In addition, the traditionally used crisp ordinal numbers may be hardly assignable to describe faithfully failure modes in classical FMEA. This work combines FMEA and a new aggregation operator for Interval-Valued Intuitionistic Fuzzy Numbers (IVIFN) to overcome both shortcomings.

Keywords

Failure mode and effect analysis (FMEA) · Interval-valued intuitionistic fuzzy number (IVIFN) · Rank reversal · Aggregation · Fuzzy consensus

1 Introduction

FMEA was first introduced by the National Aeronautics and Space Administration, known as NASA in the 1960s for the aerospace industry. It was developed as a systematic method to identify known and potential Failure Modes (FM) and their causes. To eliminate or reduce the identified risks, it proceeds by assessing the relative risk traditionally using the Risk Priority Number (RPN), which is the product of Occurrence (O), Severity (S), and Detection (D). Despite being useful in real situations, the classical FMEA method, based on crisp numbers—classically whole numbers between 1 and 10—for evaluation, suffers many drawbacks (Liu et al. 2013). Mainly, it is harder to fix a crisp assessment for risk evaluation than to use linguistic variables. Those linguistic variables are seamlessly incorporated into fuzzy logic, which can accurately model human reasoning (El Alaoui et al. 2016).

Intuitionistic Fuzzy Sets (IFS), especially Interval-Valued-IFS (IV-IFS), represent a powerful tool to deal with such situations. However, classically used aggregation operators can have a veto behavior with extreme evaluations. Thus the proposed operator aims to avoid the veto effect.

Despite the increasing research connecting FMEA and IV-IFS (Li et al. 2019; Liu et al. 2019; Tooranloo and Ayatollah 2017; Wang et al. 2016; Zhao et al. 2017), little focus has been given to consensus within the FMEA context (Li et al. 2019). Note that each decision-making process involves several Decision-Makers (DM), imposing the search for a consensus. In this work, the opinions expressed by linguistic variables are converted to their equivalent IV-IFS. Then, the metric used—based on similarity—permits iteratively allocating greater weights for coherent opinions, leading to consensual group evaluation.

The rest of this work is as follows: Sect. 2 presents the classical FMEA procedure. Section 3 discusses the shortcomings of classical FMEA. Section 4 presents the required fuzzy logic background. In Sect. 5, the proposed interval-valued intuitionistic fuzzy FMEA approach is presented. In Sect. 6, an illustrative example is given with relative comparisons. Section 7 concludes this manuscript.

M. El Alaoui (✉)
National School of Applied Sciences, Sultan Moulay Sliman University - USMS, Beni-Mellal, Morocco
e-mail: mohamedelalaoui208@gmail.com

Laboratory of Physics and Engineering Sciences, Research Team in Embedded Systems Engineering, Automation, Signal, Telecommunications and Smart Materials, Polydisciplinary Faculty, Sultan Moulay Sliman University - USMS, Beni-Mellal, Morocco

M. Gouskir et al. (eds.), *Cutting-Edge Artificial Intelligence Applications*, Sustainable Artificial Intelligence-Powered Applications,
https://doi.org/10.1007/978-3-032-00125-2_6

2 Classical FMEA Procedure

The classical FMEA follows these steps (Liu 2016):

Step 1: Determine the scope of FMEA analysis. In this step, the objectives of risk assessment as well as project boundaries and the approach to be taken are defined.

Step 2: Assemble the FMEA team. A cross-functional and multidisciplinary team is to be selected, to cover all related activities in the assessment process.

Step 3: Understand the system to be analyzed.

Step 4: Brainstorm FM of each component and their effects.

Step 5: Determine the occurrence (O), severity (S), and detectability (D) for each FM. Classically, the three risk factors O, S, and D are assessed using a notation scale from 1 to 10, as shown in Tables 1, 2, and 3.

Table 1 Occurrence rating in classical FMEA

Rating	Failure probability	Failure possible rate
1	Nearly impossible	1/150000> …
2	Remote	1/15000≥ … ≥1/150000
3	Low	1/2000≥ … ≥1/15000
4	Relatively low	1/400≥ … ≥1/2000
5	Moderate	1/80≥ … ≥1/400
6	Moderately high	1/20≥ … ≥1/80
7	High	1/8≥ … ≥1/20
8	Repeated failure	1/3≥ … ≥1/8
9	Very high	1/2≥ … ≥ 1/3
10	Extremely high	… ≥ 1/2

Table 2 Severity rating in classical FMEA

Rating	Effect
1	None
2	Very minor
3	Minor
4	Very low
5	Low
6	Moderate
7	High
8	Very high
9	Hazardous with warning
10	Hazardous without warning

Table 3 Detection rating in classical FMEA

Rating	Detection
1	Almost certain
2	Very high
3	High
4	Moderately high
5	Moderate
6	Low
7	Very low
8	Remote
9	Very remote
10	Absolutely impossible

Step 6: Calculate the RPN for each FM.

$$\mathrm{RPN} = O*S*D \tag{1}$$

Step 7: Prioritize the FM for preventive actions.

Step 8: Prepare FMEA report by summarizing the analysis results.

Step 9: Calculate the revised RPNs as the revised modes are reduced or eliminated.

3 Shortcomings of Classical FMEA

The FMEA is a useful tool for risk assessment. However, the crisp version suffers many drawbacks presented in Liu et al. (2013):

- The three risk factors O, S, and D are treated equally, which may not necessarily reflect real situations.
- The risk factors are easier to express by linguistic variables such as "likely," "important," or "high" than with crisp numbers used in classical FMEA.
- The commonly used notation scale, from 1 to 10. Can only generate 120 different numbers in the interval 1 to 1000.
- Different combinations of O, S, and D could result in the same RPN. However, the risks evaluated are very dissimilar. Two events with the respective evaluations, 2, 5, 3 and 3, 2, 5 for O, S, and D have the same RPN. However, the inducted risk could be very different (Kumru and Kumru 2013).
- The three risk factors are evaluated by discrete ordinal scales. However, the multiplication is meaningless on ordinal scales.

Many improvements have been proposed to improve the original FMEA relatively to two-tier. The first deals with the theories modeling uncertainty including the Dempster–Shafer theory (Chen and Deng 2018; Wang et al. 2019) and fuzzy logic (Chanamool and Naenna 2016; Balaraju et al. 2019). The latter tries to improve capturing and modeling human reasoning. In that sense, linguistic variables represent an easier tool for DMs than conventional numbers. In Liu et al. (2013), the authors showed that fuzzy logic is the most used approach in FMEA extensions.

4 Fuzzy Logic

The classical rating, based on crisp numbers, can distort reality. Sometimes, there is no clear difference between the interpretations of two adjacent notes. Fuzzy logic happens to be a powerful tool in dealing with vague situations, where any boundary fixed between nearby situations may be challenged.

In fuzzy logic, each element has a membership function to a certain set. Let X be a universal set and F a fuzzy subset in X, F is defined as follows (Atanassov 1986; Zadeh 1965):

$$F = \{\langle x, \mu_F(x)\rangle | x \in X\}$$

where $\mu_F(x)$ is the degree of membership of x in F in the unity interval.

$$\mu_F : X \to [0, 1].$$

Atanassov (1986) extended the notion of fuzzy set by introducing intuitionistic fuzzy sets. The last contains a membership function, a non-membership function, and a hesitancy function.

An intuitionistic fuzzy set A is defined as follows (Atanassov 1986; El Alaoui and Ben-azza 2017a):

$$A = \{\langle x, \mu_A(x), \upsilon_A(x)\rangle | x \in X\}$$

where $\mu_A(x)$ and $\upsilon_A(x)$ are, respectively, the membership function and the non-membership function, with the following conditions:

$$\mu_A : X \to [0, 1], \quad x \in X \to \mu_A(x) \in [0, 1]$$

$$\upsilon_A : X \to [0, 1], \quad x \in X \to \upsilon_A(x) \in [0, 1]$$

$$\mu_A(x) + \upsilon_A(x) \leq 1, \quad \forall x \in X.$$

The hesitancy function can be computed by the following:

$$\pi_A(x) = 1 - [\mu_A(x) + \upsilon_A(x)], \forall x \in X.$$

Fuzzy logic was made to permit fuzzy reasoning, so imposing the DM to give an exact value of membership and non-membership functions is counter to the essence of fuzzy logic. Atanassov and Gargov (1989) extended the IFS, to IV-IFS. In the last, IV-IFS, the membership and the non-membership are intervals included in the [0, 1] interval. An IV-IFS A satisfies the following:

$$A = \{\langle x, M_A(x), N_A(x)\rangle | x \in X\}$$

where $M_A(x) \subset [0, 1]$ and $N_A(x) \subset [0, 1]$ are, respectively, the membership interval and the non-membership interval, which verify the following inequality (Atanassov and Gargov 1989):

$$\sup M_A(x) + \sup N_A(x) \leq 1.$$

For convenience we note an IVIFN as $\beta = ([a, b], [c, d])$ where $a = \inf M_\beta$, $b = \max M_\beta$, $c = \inf N_\beta$, and $d = \max N_\beta$. M_β and N_β are interval numbers that verify $0 \leq a \leq b \leq 1$ and $0 \leq c \leq d \leq 1$.

The similarity between two IVIFN $\beta_1([a_1, b_1], [c_1, d_1])$ and $\beta_2([a_2, b_2], [c_2, d_2])$ can be computed as a distance dual (Santini and Jain 1997) by the following (Atanassov 1999):

$$\mathrm{Sim}(\beta_1, \beta_2) = 1 - \frac{1}{4}(|a_1 - a_2| + |b_1 - b_2| + |c_1 - c_2| + |d_1 - d_2|).$$

5 Proposed Interval-Valued Intuitionistic Fuzzy FMEA Model

The proposed method proceeds in three stages (Fig. 1). In the first, β_{ijk} individual evaluation for each failure mode $1 \leq i \leq n$ according to each risk factor $j = O, S, D$ is merged into a group consensual evaluation β_{jk} using algorithm 1. Then the three evaluations (β_{Ok}, β_{sk}, and β_{Dk}) are merged into the overall FM evaluation β_{jk} using Eqs. (2.1)–(2.4). The last is converted to its crisp equivalent using Eq. (3) to permit ranking.

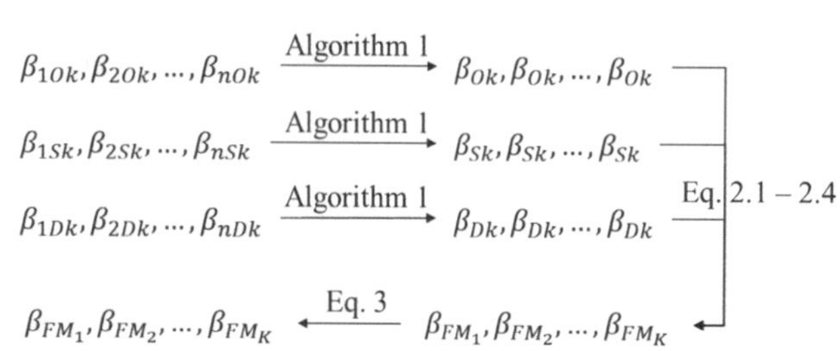

Fig. 1 Stages of the proposed interval-valued intuitionistic fuzzy FMEA model

Despite the growing literature linking FMEA and IV-IFS (Li et al. 2019; Liu et al. 2019; Tooranloo and Ayatollah 2017; Wang et al. 2016; Zhao et al. 2017), little attention has been paid to consensus in the FMEA context (Li et al. 2019). While the existing methods aim to achieve a certain threshold degree (Liu et al. 2021; Yue and Zhu 2024; Zhang et al. 2023), the following algorithm aims to minimize weighted incoherencies between expressed opinions. The incoherence $c - \mathrm{Sim}(\beta^l, \beta_i)$ is computed as a similarity inverse $\mathrm{Sim}(\beta^l, \beta_i)$ where β_i is the individual opinion and β^l the consensual collective evaluation at step l. Here, the approach proposed in El Alaoui et al. (2018) to derive optimal weight for each decision is adapted to the FMEA context as described in Algorithm 1.

Algorithm 1

Step 1: Each DM $1 \le i \le n$ assesses each FM using a linguistic variables.

Step 2: Convert the linguistic variable to their equivalent IVIFN $\beta_i = \left(\left[a_{\beta_i}, b_{\beta_i}\right], \left[c_{\beta_i}, d_{\beta_i}\right]\right)$.

Step 3: Set the initial aggregation weights such that $0 \le w_i^{(0)} \le 1$ and $\sum_{i=1}^{n} w_i = 1$. The iterations are labeled $l = 0, 1, 2, \ldots$.

Step 4: Compute the aggregated consensus $\beta^l\left(\left[a^l, b^l\right], \left[c^l, d^l\right]\right)$ at step l by

$$a^l = 1 - \prod_{i=1}^{n} (1 - a_i)^{w_i^{(l)}}$$

$$b^l = 1 - \prod_{i=1}^{n} (1 - b_i)^{w_i^{(l)}}$$

$$c^l = \prod_{i=1}^{n} c_i^{w_i^{(l)}}$$

$$d^l = \prod_{i=1}^{n} d_i^{w_i^{(l)}}$$

Step 5: Let $W^l = \left(w_1^{(l)}, w_2^{(l)}, \ldots, w_n^{(l)}\right)$. Compute W^{l+1} as follows:

$$w_i^{(l+1)} = \frac{\left(1/\left(c - \mathrm{Sim}\left(\beta^l, \beta_i\right)\right)\right)^{1/(m-1)}}{\sum_{j=1}^{n}\left(1/\left(c - \mathrm{Sim}\left(\beta^l, \beta_j\right)\right)\right)^{1/(m-1)}}$$

Step 6: If $\left\| W^{l+1} - W^l \right\| > \varepsilon$, set $l = l + 1$ and go to step 4. Else Stop.

The relevance of the selected operators was demonstrated experimentally in El Alaoui et al. (2019).

Let $\beta_{Ok}([a_{Ok}, b_{Ok}], [c_{Ok}, d_{Ok}])$, $\beta_{Sk}([a_{Sk}, b_{Sk}], [c_{Sk}, d_{Sk}])$, and $\beta_{Dk}([a_{Dk}, b_{Dk}], [c_{Dk}, d_{Dk}])$ the aggregated occurrence, severity, and detectability evaluation for the kth FM, respectively. And w_O, w_S, and w_D the respective weights for the three risk factors. The overall evaluation for the kth FM $\beta_k([a_k, b_k], [c_k, d_k])$ can be computed as follows (El Alaoui and Ben-azza 2017a, 2017b):

$$a_k = 1 - \frac{\prod_j^{O,S,D} (2 - w_j - a_{jk})^{\frac{1}{3}} - \prod_j^{O,S,D} (1 - w_j)^{\frac{1}{3}}}{\prod_j^{O,S,D} (2 - w_j)^{\frac{1}{3}} - \prod_j^{O,S,D} (1 - w_j)^{\frac{1}{3}}} \quad (2.1)$$

$$b_k = 1 - \frac{\prod_j^{O,S,D} (2 - w_j - b_{jk})^{\frac{1}{3}} - \prod_j^{O,S,D} (1 - w_j)^{\frac{1}{3}}}{\prod_j^{O,S,D} (2 - w_j)^{\frac{1}{3}} - \prod_j^{O,S,D} (1 - w_j)^{\frac{1}{3}}} \quad (2.2)$$

$$c_k = \frac{\prod_j^{O,S,D} (1 - w_j + c_{jk})^{\frac{1}{3}} - \prod_j^{O,S,D} (1 - w_j)^{\frac{1}{3}}}{\prod_j^{O,S,D} (2 - w_j)^{\frac{1}{3}} - \prod_j^{O,S,D} (1 - w_j)^{\frac{1}{3}}} \quad (2.3)$$

$$d_k = \frac{\prod_j^{O,S,D} (1 - w_j + d_{jk})^{\frac{1}{3}} - \prod_j^{O,S,D} (1 - w_j)^{\frac{1}{3}}}{\prod_j^{O,S,D} (2 - w_j)^{\frac{1}{3}} - \prod_j^{O,S,D} (1 - w_j)^{\frac{1}{3}}} \quad (2.4)$$

It should be mentioned that the aggregation proposed (Eqs. (2.1)–(2.4)) unlike step 4 in algorithm 1 aims to avoid the veto behavior of extreme evaluations. A single null c_i or d_i will force the aggregated outcome to nullity despite the other aggregated opinions in algorithm 1. Similarly, a single a_i or b_i that equals 1 will have a veto behavior, which explains the choice of equivalent IVIFNs (Table 4). In that sense the aggregation (Eqs. (2.1)–(2.4)) avoids these shortcomings. Furthermore, several methods chose carefully the used

Table 4 Linguistic variables for failure modes

Linguistic variable	Detection
Extremely low (EL)	([0, 0.05], [0.9, 0.9])
Very low (VL)	([0.05, 0.1], [0.8, 0.9])
Low (L)	([0.1, 0.2], [0.7, 0.8])
Medium low (ML)	([0.3, 0.4], [0.5, 0.6])
Medium (M)	([0.5, 0.5], [0.5, 0.5])
Medium high (MH)	([0.5, 0.6], [0.3, 0.4])
High (H)	([0.7, 0.8], [0.1, 0.2])
Very high (VH)	([0.8, 0.9], [0.05, 0.1])
Extremely high (EH)	([0.9, 0.9], [0, 0.05])

IVIFNs to avoid the veto behavior of the aggregation operators. Authors in Liu et al. (2019); Huang and Xiao 2021) opted for ([0.99, 0.99], [0.01, 0.01]) as the equivalent IVIFN for an exceptionally high evaluation. While in the proposed (Table 4), extrema value (0 and 1) are tolerated. The value 1 was not used just to fit the comparison.

Authors in Mishra and Kumar (2021) address intuitionistic fuzzy aggregation operators among other fuzzy aggregation operators and pointed that the weighted geometric aggregation operators and the weighted averaging aggregation operators suffer several drawbacks including the veto behavior of extrema values.

To rank IVIFNs, the method proposed in Nayagam and Sivaraman (2011) is adopted. Let $\beta = ([a, b], [c, d])$ be an IVIFN. The accuracy function is computed as follows:

$$LG(\beta) = \frac{a + b + \delta * (2 - a - b - c - d)}{2} \quad (3)$$

where $0 \leq \delta \leq 1$ presents the aversion degree of the DM. An optimistic DM will choose $\delta = 1$ a pessimistic DM will fix $\delta = 0$. Generally, we choose $\delta = 0.5$.

Hence, having the list of FMs in step 4 in the classical FMEA algorithm, steps from 5 to 7 in classical FMEA will be modified by the following:

Step 5′: Determine occurrence, severity, and detectability for each FM according to Table 4. Aggregate the evaluation to have a consensual assessment by Algorithm 1.

Step 6′: Merge consensual evaluation for each FM in accordance with the three attributes into an overall evaluation for each FM using (Eqs. (2.1), (2.2), (2.3), and (2.4)).

Step 7′: Rank FMs using (Eq. (3)).

Other steps remain unchanged.

To ease comparison the same linguistic variables as in Zhao et al. (2017) are adopted (Table 4).

6 Application Example

Here the example treated in Zhao et al. (2017) is adopted, in which three DMs evaluate steel production FMs. Table 5 resumes the studied FMs, their causes, and their effects. Table 6 contains the linguistic assessment delivered by three DMs.

Using Algorithm 1, the group evaluations are (Tables 7 and 8):

Fixing the same risk weights (0.4, 0.274, 0.326) as in Zhao et al. (2017), the aggregated evaluations using Eqs. (2.1), (2.2), (2.3) and (2.4) are:

Zhao et al. method (2017) fixes DMs weights in all situations, favoring inconsistent opinions, which can be observed surrounding FM_2 and FM_{10}, topics of contention.

Li et al. (2019) advocate the search for consensus among DMs before the aggregation under the three risk factors which results in an extremely high detectability risk for FM_2 with a significant difference in favor of FM_{10} according to other criteria. All the more, the weight assigned to detectability in their method happens to be the greatest. However, the FM_2 score recedes to the fifth position, absorbed by a small result according to the occurrence, which calls into question the classical multiplicative aggregation operator. Similar reproaches can be formulated with respect to the method in Liu et al. (2019), surrounding the intermediate assessed FMs.

Table 5 FMEA of a steel production process

Index	Failure modes	Causes	Effects
FM_1	Non-acceptable formation	Non-conductive scrap	Electrode may break
FM_2	Nipple thread pitted	Proper coverage not obtained	Electrode comes out
FM_3	Arc formation loss	Leakage of water, proper gripping loss	Breakage of electrode, busting of bus tube
FM_4	Burn-out electrode	Cooler not working properly	Missing of electrode
FM_5	Breaking of house pipe	Wearing of pipe due to use	Emulsion lost, pressure decrease
FM_6	Problem in arm movement	Severe leakage	Arc formation problem
FM_7	Refractory damage	Because of slag	Heat convection takes place
FM_8	Formation of steam	Roof leak	Hot gas comes out
FM_9	Refractory line damage	By hot gas	Heat convection takes place
FM_{10}	Movement of roof stop	Jam of plunger in unloader valve	Roof cannot move

Table 6 Linguistic evaluation

Index	Occurrence			Severity			Detectability		
	DM_1	DM_2	DM_3	DM_1	DM_2	DM_3	DM_1	DM_2	DM_3
FM_1	L	EL	L	H	VH	L	L	ML	L
FM_2	L	EL	L	H	H	MH	VH	H	VH
FM_3	EL	EL	VL	VH	EH	H	L	ML	ML
FM_4	EL	EL	VL	ML	ML	ML	L	EL	VL
FM_5	L	L	L	ML	ML	ML	H	VH	MH
FM_6	L	L	L	ML	ML	ML	H	VH	MH
FM_7	L	L	L	ML	M	MH	L	EL	L
FM_8	EL	EL	VL	VH	EH	H	ML	ML	ML
FM_9	L	L	L	ML	ML	ML	H	VH	MH
FM_{10}	L	ML	ML	H	VH	MH	H	ML	ML

Table 7 Consensual evaluation using the proposed approach

Index	Occurrence	Severity	Detectability
FM_1	([0.068, 0.153], [0.761, 0.832])	([0.068, 0.153], [0.761, 0.832])	([0.068, 0.153], [0.761, 0.832])
FM_2	([0.636, 0.760], [0.142, 0.240])	([0.636, 0.760], [0.142, 0.240])	([0.636, 0.760], [0.142, 0.240])
FM_3	([0.172, 0.273], [0.626, 0.727])	([0.172, 0.273], [0.626, 0.727])	([0.172, 0.273], [0.626, 0.727])
FM_4	([0.068, 0.153], [0.761, 0.832])	([0.068, 0.153], [0.761, 0.832])	([0.068, 0.153], [0.761, 0.832])
FM_5	([0.645, 0.748], [0.144, 0.252])	([0.645, 0.748], [0.144, 0.252])	([0.645, 0.748], [0.144, 0.252])
FM_6	([0.771, 0.874], [0.063, 0.126])	([0.771, 0.874], [0.063, 0.126])	([0.771, 0.874], [0.063, 0.126])
FM_7	([0.017, 0.067], [0.865, 0.9])	([0.017, 0.067], [0.865, 0.9])	([0.017, 0.067], [0.865, 0.9])
FM_8	([0.818, 0.874], [0, 0.1])	([0.818, 0.874], [0, 0.1])	([0.818, 0.874], [0, 0.1])
FM_9	([0.239, 0.34], [0.559, 0.66])	([0.239, 0.34], [0.559, 0.66])	([0.239, 0.34], [0.559, 0.66])
FM_{10}	([0.017, 0.067], [0.865, 0.9])	([0.017, 0.067], [0.865, 0.9])	([0.017, 0.067], [0.865, 0.9])

Note that the utility value in methods (Li et al. 2019; Zhao et al. 2017) ranks FM_2 in the first position, while the final ranking indices achieve different order, questioning the rank reversal in these methods.

Concerning FMs, it seems that the biggest rank shifts happen with regard to FMs that are highly evaluated according to certain criteria and low regarding others.

Table 8 Proposed ranking and comparison

Index	Overall	Ranking (Zhao et al. 2017)	Ranking (Li et al. 2019)	Ranking (Liu et al. 2019)	Ranking proposed
FM_1	([0.307, 0.414], [0.488, 0.581])	8	6	8	8
FM_2	([0.522, 0.623], [0.291, 0.373])	2	5	1	1
FM_3	([0.393, 0.462], [0.43, 0.517])	4	8	3	7
FM_4	([0.124, 0.198], [0.717, 0.786])	10	10	10	10
FM_5	([0.384, 0.49], [0.414, 0.51])	5	2	5	4
FM_6	([0.384, 0.49], [0.414, 0.51])	5	2	5	4
FM_7	([0.208, 0.291], [0.623, 0.704])	9	9	9	9
FM_8	([0.411, 0.48], [0.412, 0.499])	3	7	2	3
FM_9	([0.384, 0.49], [0.414, 0.51])	5	2	5	4
FM_{10}	([0.473, 0.582], [0.315, 0.418])	1	1	4	2

Note first that all methods agreed on the 10th and 9th ranked FM, respectively, FM_4 and FM_7

7 Conclusion

This work presented an extended FMEA method in which decision-makers' assessments were presented by interval-valued intuitionistic fuzzy numbers. A consensus is derived using a minimization process. Then an extended multiplicative aggregation operator is used to merge collective decisions avoiding the classical drawbacks of traditionally used aggregation operators. A successful illustrative example was presented. Further extensions could include other uncertainty theories.

References

Atanassov KT (1986) Intuitionistic fuzzy sets. Fuzzy Sets Syst 20:87–96. https://doi.org/10.1016/S0165-0114(86)80034-3

Atanassov KT, Gargov G (1989) Interval valued intuitionistic fuzzy sets. Fuzzy Sets Syst 31:343–349. https://doi.org/10.1016/0165-0114(89)90205-4

Atanassov KT (1999) Intuitionistic Fuzzy Sets: Theory and Applications. Physica-Verlag Heidelberg

Balaraju J, Govinda Raj M, Murthy CS (2019) Fuzzy-FMEA risk evaluation approach for LHD machine – A case study. J Sustain Min 18:257–268. https://doi.org/10.1016/j.jsm.2019.08.002

Chanamool N, Naenna T (2016) Fuzzy FMEA application to improve decision-making process in an emergency department. Appl Soft Comput 43:441–453. https://doi.org/10.1016/j.asoc.2016.01.007

Chen L, Deng Y (2018) A new failure mode and effects analysis model using Dempster–Shafer evidence theory and grey relational projection method. Eng Appl Artif Intell 76:13–20. https://doi.org/10.1016/j.engappai.2018.08.010

El Alaoui M, Ben-azza H, El Yassini K (2018) Optimal weighting method for interval-valued intuitionistic fuzzy opinions. Notes Intuitionistic Fuzzy Sets 24:106–110. https://doi.org/10.7546/nifs.2018.24.3.106-110

El Alaoui M, Ben-azza H, El Yassini K (2019) Achieving consensus in interval valued intuitionistic fuzzy environment. Procedia Comput Sci 148:218–225. https://doi.org/10.1016/j.procs.2019.01.064

El Alaoui M, Ben-azza H (2017) Generalization of the weighted product aggregation applied to data fusion of intuitionistic fuzzy quantities. In: 2017 Intelligent Systems and Computer Vision (ISCV). pp 1–6

El Alaoui M, Ben-azza H (2017) Aggregation of performance indicators for supply chain and fuzzy logic extensions applied to green supply chain. In: 2017 International Colloquium on Logistics and Supply Chain Management (LOGISTIQUA). pp 36–41

El Alaoui M, Ben-azza H, Zahi A (2016) New Multi-criteria Decision-Making Based on Fuzzy Similarity, Distance and Ranking. In: Proceedings of the Third International Afro-European Conference for Industrial Advancement — AECIA 2016. Springer, Cham, pp 138–148

Huang G, Xiao L (2021) Failure mode and effect analysis: An interval-valued intuitionistic fuzzy cloud theory-based method. Appl Soft Comput 98:106834. https://doi.org/10.1016/j.asoc.2020.106834

Kumru M, Kumru PY (2013) Fuzzy FMEA application to improve purchasing process in a public hospital. Appl Soft Comput 13:721–733. https://doi.org/10.1016/j.asoc.2012.08.007

Li Y-L, Wang R, Chin K-S (2019) New failure mode and effect analysis approach considering consensus under interval-valued intuitionistic fuzzy environment. Soft Comput. https://doi.org/10.1007/s00500-018-03706-5

Liu H-C, Liu L, Liu N (2013) Risk evaluation approaches in failure mode and effects analysis: A literature review. Expert Syst Appl 40:828–838. https://doi.org/10.1016/j.eswa.2012.08.010

Liu H-C, You J-X, Duan C-Y (2019) An integrated approach for failure mode and effect analysis under interval-valued intuitionistic fuzzy environment. Int J Prod Econ 207:163–172. https://doi.org/10.1016/j.ijpe.2017.03.008

Liu P, Zhang X, Pedrycz W (2021) A consensus model for hesitant fuzzy linguistic group decision-making in the framework of Dempster–Shafer evidence theory. Knowl-Based Syst 212:106559. https://doi.org/10.1016/j.knosys.2020.106559

Liu H-C (2016) FMEA Using Uncertainty Theories and MCDM Methods. Springer Singapore

Mishra A, Kumar A (2021) Aggregation Operators for Various Extensions of Fuzzy Set and Its Applications in Transportation Problems. Springer, Singapore

Nayagam VLG, Sivaraman G (2011) Ranking of interval-valued intuitionistic fuzzy sets. Appl Soft Comput 11:3368–3372. https://doi.org/10.1016/j.asoc.2011.01.008

Santini S, Jain R (1997) Similarity is a Geometer. Multimed Tools Appl 5:277–306. https://doi.org/10.1023/A:1009651725256

Tooranloo HS, Ayatollah AS (2017) Pathology the Internet Banking Service Quality Using Failure Mode and Effect Analysis in Interval-Valued Intuitionistic Fuzzy Environment. Int J Fuzzy Syst 19:109–123. https://doi.org/10.1007/s40815-016-0265-y

Wang L-E, Liu H-C, Quan M-Y (2016) Evaluating the risk of failure modes with a hybrid MCDM model under interval-valued intuitionistic fuzzy environments. Comput Ind Eng 102:175–185. https://doi.org/10.1016/j.cie.2016.11.003

Wang H, Deng X, Zhang Z, Jiang W (2019) A New Failure Mode and Effects Analysis Method Based on Dempster–Shafer Theory by Integrating Evidential Network. IEEE Access 7:79579–79591. https://doi.org/10.1109/ACCESS.2019.2923064

Yue L, Zhu J (2024) A New FMEA Approach with a Consensus Reaching Model Considering Bounded Confidences Based on Deep Learning Classification Algorithm. In: 2024 IEEE 2nd International Conference on Control, Electronics and Computer Technology (ICCECT). pp 484–486

Zadeh LA (1965) Fuzzy sets. Inf Control 8:338–353. https://doi.org/10.1016/S0019-9958(65)90241-X

Zhang H, Liu S, Dong Y, Chiclana F, Herrera-Viedma EE (2023) A Minimum Cost Consensus-Based Failure Mode and Effect Analysis Framework Considering Experts' Limited Compromise and Tolerance Behaviors. IEEE Trans Cybern 53:6612–6625. https://doi.org/10.1109/TCYB.2022.3212093

Zhao H, You J-X, Liu H-C (2017) Failure mode and effect analysis using MULTIMOORA method with continuous weighted entropy under interval-valued intuitionistic fuzzy environment. Soft Comput 21:5355–5367. https://doi.org/10.1007/s00500-016-2118-x

The Influence of Artificial Intelligence on Knowledge Management in Small and Medium Enterprises

Meryem Samine, Zakaria Khaouch, and Nourreeddine Kouider

Abstract

Artificial Intelligence (AI) becomes continuously more implanted in various forms in business processes and organizational activities, revolutionizing the way organizations operate and manage their operations. One of the most significant impacts of AI is its integration in knowledge management that holds a vast potential by leading significant advancements in the way organizations collect, store, share, and strategically manage information to achieve their business objectives. In this context, the purpose of this work is to examine the impact of artificial intelligence on knowledge management and its elements in small- and medium-sized enterprises in Morocco. The study uses quantitative research, where 200 responses are collected to gather the required data for testing the hypotheses of the study. The results have revealed that adopting AI in small and medium enterprises is significant for the organizations to accelerate the adoption of a knowledge management solution, improve the connection of employees, and give them access to necessary resources to perform their tasks efficiently. Finally, it is noted that knowledge management system enhanced by artificial intelligence is the most effective solution.

Keywords

Small and medium enterprises (SMEs) · Knowledge management · Artificial intelligence

M. Samine (✉) · Z. Khaouch · N. Kouider
Industrial and Surface Engineering, Faculty of Science and Technology, Sultan Moulay Slimane University, Beni Mellal, Morocco
e-mail: meryem.samine@usms.ma

1 Introduction

Small- and medium-sized enterprises (SMEs) are universally recognized as the fundamental pillars of a dynamic and sustainable economy. Their role extends far beyond improving the entrepreneurial ecosystem; they are key drivers of employment opportunities, innovation, foster competition, and expand markets. SMEs help in promoting economic flexibility, sustainability, and inclusivity, supporting long-term prosperity and economic well-being of individuals. The evolution of this portion of businesses is related to many factors. One of the most practical ways for competing in the national and international markets is by adopting a methodology to convert information or data into knowledge. Most employees often leaving with their skills, ideas, and taking all their expertise with them. Thus, companies need to adapt their approach to knowledge management (KM).

The advantages of knowledge management are widely recognized, demonstrated by its successful adoption in many large companies. For this reason, KM is considered an important improvement tool and requires the appropriate system. In this sense, the use of artificial intelligence (AI) improves the efficiency and accuracy of KM by automating tasks, organizing, analyzing, and managing all of that data. Essentially, preventing and identifying issues that may not be readily apparent to humans.

Firstly, many studies have shown the positive impact of KM on SME. Coyte et al. (2012) examined strategy and processes used in an Australian SME to control the management of knowledge resources. Alegre et al. (2013) investigated the influence of knowledge management (KM) on innovation performance in biotechnology companies. According to Harris et al. (2013), knowledge management has opened up new opportunities for small- and medium-sized enterprises (SMEs) in peripheral regions to use knowledge as a means of gaining a competitive advantage. Hussain et al. (2019) analyzed the impact of various knowledge

M. Gouskir et al. (eds.), *Cutting-Edge Artificial Intelligence Applications*, Sustainable Artificial Intelligence-Powered Applications,
https://doi.org/10.1007/978-3-032-00125-2_7

management processes on innovation capacities, revealing the benefits of implementing effective knowledge management systems in SMEs. Zyl et al. (2022) recommended the adoption of knowledge management system to facilitate the integration of technology in SMEs. Similarly, Cheng et al. (2024) noted that encouraging knowledge management and organizational learning enables SMEs to successfully adopt Industry 4.0, ensuring alignment with global industrial development. Akbar et al. (2024) highlight the direct influence of KM and innovation on SME performance in Yogyakarta. Trisa Diptami and Budi (2025) demonstrated that knowledge management has a positive impact on both strategic flexibility and organizational creativity in SMEs.

Secondly, many papers studied the role of information systems especially AI in the knowledge management activities of SMEs. Gresty (2013) noted that information systems have been seen as essential tools in the facilitation of knowledge. Nevertheless, it has also been observed that SMEs may face challenges from having too many competing knowledge management systems, so they can face risks by depending too heavily on these systems. Alvarez et al. (2016) affirmed that smaller firms tend to use collaborative tools for knowledge sharing, while larger firms focus more on IT tools to support knowledge management. It was discovered by Cerchione and Esposito (2017) that SMEs favor traditional approaches like problem-solving, teamwork, and learning through experience, instead of adopting special KM practices such as communities of practice, data mining, and knowledge mapping. According to Iyelolu et al. (2024a), AI technologies have the potential to greatly boost customer engagement, improve the efficiency of operations, and create a competitive advantage for SMEs. Akoh (2024) developed a framework that promotes AI adoption in South African manufacturing SMEs, improving their innovation and productivity. Iyelolu et al. (Iyelolu et al. 2024b) highlight that AI can revolutionize SMEs by facilitating rapid and efficient innovation. Alonta et al. (2025) indicated that AI is crucial in improving communication and digital literacy skills among SME operators in Anambra State, Nigeria.

This article is organized as follows: first, the introduction to highlight the importance of the topic and set the context. Next, background of KM notion and processes. Following this, the methodology section, which describes the data collection sources and analysis methods employed. Results are presented next, which are analyzed in the discussion section. The article wraps up with a conclusion.

2 Materials and Methods

2.1 Review of Scientific Literature

2.1.1 Knowledge Management

Researchers have presented different views and definitions of knowledge management. Nevertheless, the common consensus of KM is the best use of both individual and organizational knowledge resources. Knowledge management (KM) can be defined as a developing set of organizational structures, processes, applications, and technologies that empowers knowledge workers to enhance their creativity (Gurteen 1998). Knowledge management can also be seen from a human-focused perspective by the appreciation of employees' knowledge (Bennet and Bennet 2004). As indicated by Abbas and Sağsan (2019), knowledge management has a significant impact on green innovation and corporate sustainable development (CSD). Additionally, KM is considered equally essential for firms of all sizes, both in services and manufacturing sectors. Knowledge is an intangible concept, it manifests in two main forms: explicit knowledge and tacit knowledge (Bolisani et al. 2018). Tacit knowledge is the kind of knowledge we instinctively use but find hard to have consciousness of or to explain in words (Hadjimichael and Tsoukas 2019). Explicit knowledge or "know-what" is knowledge that is clearly expressed and documented, often found in digital formats or printed. It is generally based on standardized processes (Smith 2001).

Knowledge management was also defined as a dynamic integration of specific processes, which include the creation, storage/retrieval, transfer, and application of knowledge (Alavi and Leidner 2001); conversion-oriented knowledge, application-based processes, and protection processes (Gold et al. 2001); knowledge generation, capture, sharing, and utilization (Zaim et al. 2018); knowledge creation, knowledge storage, knowledge transfer, and knowledge applications (Al Shraah et al. 2022). In this study, based on the most cited KM processes, in the scientific literature, KM is described as a dynamic combination of particular processes, including knowledge creation, knowledge storage, knowledge sharing, and knowledge application (Fig. 1).

2.1.2 Artificial Intelligence

Artificial intelligence is the use of machines, especially computer systems to imitate human intelligence processes. Jarrahi (2018) described the term as a set of tools, strategies, and practices employed inside an organizational context

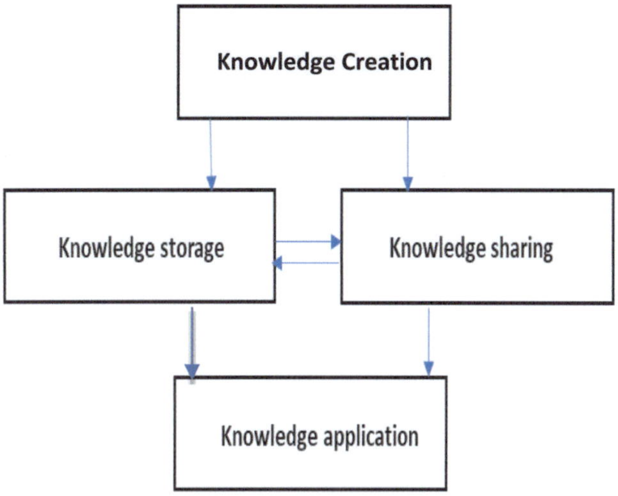

Fig. 1 Knowledge management processes

to support the firm. Bellomarini et al. (2019) underscore that adopting a robust AI strategy is crucial for modern knowledge-driven companies. They proposed knowledge graphs as the reference technology for the enterprise AI applications.

According to Ghosh and Arunachalam (2021), artificial intelligence and its associated technologies, such as machine learning, deep learning, and natural language processing, have made a powerful impact in the word. Their influence is effective, touching nearly every field from agriculture to education and even to the space technology. Machine learning is a branch of artificial intelligence that analyzes patterns in large datasets and uses algorithms to generate insights and recommendations to support decision-making (Hussain et al. 2020). Deep learning (DL), a subfield of machine learning (ML) and artificial intelligence (AI), is now viewed as foundational technology in today's Fourth Industrial Revolution (Sarker 2021). Natural Language Processing (NLP) is a fundamental discipline in artificial intelligence and computer science. It involves theories and techniques that enable interaction between humans and computers through natural language. As an interdisciplinary field, NLP aims to transform human language into executable commands for computers (Kang et al. 2020).

2.2 Hypotheses Development

Conforming to literature of the study, it is proposed that artificial intelligence has positive impact on the chosen processes of knowledge management in small- and medium-sized organizations in Morocco. By using chi-square test of independence, we will determine whether AI and KM are likely to be related or not. Current research has conceptualized subsequent hypotheses (Fig. 2).

The null hypothesis is written as follows:

H0: Artificial intelligence and knowledge management are independent.

H01: Artificial intelligence and knowledge creation are independent.

H02: Artificial intelligence and knowledge storage are independent.

H03: Artificial intelligence and knowledge sharing are independent.

H04: Artificial intelligence and knowledge application are independent.

The alternative hypothesis is the opposite.

H1: Artificial intelligence and knowledge management are not independent.

Fig. 2 Model of the study

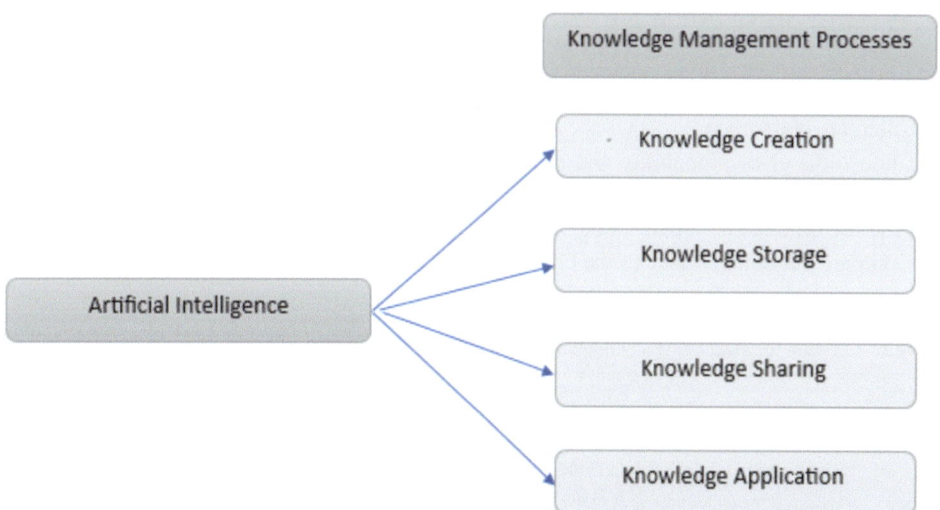

H11: Artificial intelligence and knowledge creation are not independent.

H12: Artificial intelligence and knowledge storage are not independent.

H13: Artificial intelligence and knowledge sharing are not independent.

H14: Artificial intelligence and knowledge application are not independent.

3 Results and Discussion

This work is based on a survey to study the impact of artificial intelligence on knowledge management. It was distributed to industrial people in Moroccan small and medium enterprises by sending emails and direct contact. The survey included a variety of question formats, open questions, multiple-choice questions, and in certain types of answers, we used Likert scale from 1 to 5. All 200 respondents of this survey worked in small or medium companies that had integrated knowledge management or interested by it. This condition guaranteed that people in the companies sampled for the questionnaire had a substantial understanding and experience in the field of knowledge management. In this research, we applied the chi-square test of independence for testing our hypotheses (Table 1).

The objective of this study was to examine the effect of artificial intelligence on knowledge management in small and medium enterprises. Table 2 illustrates the classification of analyzed enterprises based on their size (several people responded to the questionnaire at each company).

Based on the data obtained from the participants, we verified this hypothesis: we posit that there is a statistical relationship between AI and KM in Moroccan SMEs. We employed data obtained from respondents who actively implemented the KM to verify this hypothesis. By using Pearson's chi-square test of independence, we follow the steps given below:

- Calculate the expected frequencies.
- Calculate chi-square.
- Choose the error probability and calculate the degree of freedom.
- Find the critical chi-square value.
- Compare chi-square value to the critical value (Table 3).

Table 1 Formula for calculating a chi-square

Formula	Explanatory note
$X^2 = \sum \frac{(f_e - f_t)^2}{f_t}$	X^2 is the chi-square test statistic f_e is the observed frequency f_t is the expected frequency

Table 2 The arrangement of enterprises based on their size

	Number of employees	Number of enterprises
Enterprises size	1–9 employees	12
	10–49 employees	18
	50–249 employees	20

Table 3 Results of testing the first hypothesis

Pearson's chi-square test of independence			
Calculated value	Degree of freedom	Error probability	Critical value
$X^2 = 13.333$	$D_f = (4-1)(2-1) = 3$	$\alpha = 0.05$	$C_v = 7{,}815$

The X^2 value is greater than the critical value, so based on the results of our analysis using Pearson's chi-square test of independence we can assert that there exists a statistically significant relationship between the knowledge management processes (the four examined in this study) and artificial intelligence. Thus, our hypothesis is accepted.

4 Conclusion

The objective of this research was to examine how the adoption of Artificial Intelligence (AI) influences the integration of knowledge management in Moroccan small and medium enterprises. In this paper, based on the literature review we examined four important KM processes (knowledge creation, knowledge storage, knowledge sharing, and knowledge application). Thus, we highlight the need of companies to use AI because this will directly influence the integration of KM and consequently their performance.

References

Abbas, J., and Sağsan, M. (2019). Impact of knowledge management practices on green innovation and corporate sustainable development: A structural analysis. *Journal of cleaner production,* volume 229, pages 611–620.

Akbar, R. E., & Hartono, A. (2024). The Impact of Open Innovation and Knowledge Management on SME Performance in Yogyakarta. *EKOMBIS REVIEW: Jurnal Ilmiah Ekonomi dan Bisnis, 12*(3), 2633–2644.

Akoh, E. I. (2024). Adoption of artificial intelligence for manufacturing SMEs' growth and survival in South Africa: A systematic literature review. *International Journal of Research in Business and Social Science, 13*(6), 23–37.

Al Shraah, A., Abu-Rumman, A., Al Madi, F., Alhammad, F. A. F., & AlJboor, A. A. (2022). The impact of quality management practices on knowledge management processes: a study of a social security corporation in Jordan. *The TQM Journal, 34*(4), 605–626.

Alavi, M., & Leidner, D. E. (2001). Knowledge management and knowledge management systems: Conceptual foundations and research issues. *MIS quarterly*, 107–136.

Alegre, J., Sengupta, K. and Lapiedra, R. (2013), "Knowledge management and innovation performance in a high-tech SMEs industry", *International Small Business Journal*, Vol. 31 No. 4, pp. 454–470.

Alonta, G. C., Orabueze, M. C., & Onyechi, I. F. (2025). Influence of artificial intelligence on enhancing communication and digital literacy skills among SME operators in Anambra state, Nigeria. *Nnadiebube Journal of Education in Africa*, 10(1).

Alvarez, I., Zamanillo, I. and Cilleruelo, E. (2016), "Have information technologies evolved towards accommodation of knowledge management needs in Basque SMEs?", *Technology in Society*, Vol. 46, pp. 126–131.

Bellomarini, L., Fakhoury, D., Gottlob, G., & Sallinger, E. (2019, April). Knowledge graphs and enterprise AI: the promise of an enabling technology. In *2019 IEEE 35th international conference on data engineering (ICDE)* (pp. 26–37). IEEE.

Bennet, D. and Bennet, A. (2004), "The rise of the knowledge organization", in Holsapple, C.W. (Ed.), Handbook on Knowledge Management, 1. *International Handbooks on Information Systems*, Springer, Berlin, Heidelberg, Vol. 1.

Bolisani, E., Bratianu, C., Bolisani, E., & Bratianu, C. (2018). The elusive definition of knowledge. *Emergent knowledge strategies: Strategic thinking in knowledge management*, 1–22.

Cerchione, R. and Esposito, E. (2017), "Using knowledge management systems: a taxonomy of SME strategies", *International Journal of Information Management*, Vol. 37 No. 1, pp. 1551–1562.

Cheng, C. H., Li, M. H., Tang, B. J., & Cheng, Y. R. (2024). The Impact of Knowledge Management and Organizational Learning Promotion in Small and Medium Enterprises on the Implementation of Industry 4.0 and Competitiveness. *Administrative Sciences*, *14*(8), 161.

Coyte, R., Ricceri, F. and Guthrie, J. (2012), "The management of knowledge resources in SMEs: an Australian case study", *Journal of Knowledge Management*, Vol. 16 No. 5, pp. 789–807.

Moumita Ghosh and Thirugnanam Arunachalam, 2021. Introduction to Artificial Intelligence.

Gold, A. H., Malhotra, A., & Segars, A. H. (2001). Knowledge management: An organizational capabilities perspective. *Journal of management information systems*, *18*(1), 185–214.

Gresty, M. (2013), "What role do information systems play in the knowledge management activities of SMEs?", *Business Information Review*, Vol. 30 No. 3, pp. 144–151.

Gurteen, D. (1998), "Knowledge, creativity and innovation", *Journal of Knowledge Management,* Vol. 2 No. 1, pp. 5–13.

Hadjimichael, D., & Tsoukas, H. (2019). Toward a better understanding of tacit knowledge in organizations: Taking stock and moving forward. *Academy of Management Annals*, *13*(2), 672–703.

Harris, R., McAdam, R., McCausl, I. and Reid, R. (2013), "Knowledge management as a source of innovation and competitive advantage for SMEs in peripheral regions", *International Journal of Entrepreneurship and Innovation,* Vol. 14 No. 1, pp. 49–61.

Hussain, I., Qurashi, A., Mujtaba, G., Waseem, M.A. and Iqbal, Z. (2019), "Knowledge management: a roadmap for innovation in SMEs' sector of Azad Jammu & Kashmir", *Journal of Global Entrepreneurship Research*, Vol. 9 No. 1, pp. 1–18.

Hussain, F., Hussain, R., Hassan, S. A., & Hossain, E. (2020). Machine learning in IoT security: Current solutions and future challenges. *IEEE Communications Surveys & Tutorials*, *22*(3), 1686–1721.

Iyelolu, T. V., Agu, E. E., Idemudia, C., & Ijomah, T. I. (2024b). Driving SME innovation with AI solutions: overcoming adoption barriers and future growth opportunities. *International Journal of Science and Technology Research Archive*, *7*(1), 036–054.

Iyelolu, T. V., Agu, E. E., Idemudia, C., & Ijomah, T. I. (2024a). Improving Customer Engagement and CRM for SMEs with AI Driven Solutions and Future Enhancements. *International Journal of Engineering Research and Development*, *20*(8).

Jarrahi, M. H. (2018). Artificial intelligence and the future of work: Human-AI symbiosis in organizational decision making. *Business Horizons*, 61(4), 577–586. https://doi.org/10.1016/j.bushor.2018.03.007.

Kang, Y., Cai, Z., Tan, C. W., Huang, Q., & Liu, H. (2020). Natural language processing (NLP) in management research: A literature review. *Journal of Management Analytics*, *7*(2), 139–172.

Sarker, I. H. (2021). Deep learning: a comprehensive overview on techniques, taxonomy, applications and research directions. *SN computer science*, *2*(6), 1–20.

Smith, E. A. (2001). The role of tacit and explicit knowledge in the workplace. *Journal of knowledge Management*, *5*(4), 311–321.

Trisa Diptami, P., & Budi, W. S. (2025). Enhancing Innovation Performance by Utilizing Knowledge Management, Strategic Flexibility, and Organizational Creativity.

Van Zyl, W. R., Henning, S., & van der Poll, J. A. (2022). A framework for knowledge management system adoption in small and medium enterprises. *Computers*, *11*(9), 128.

Zaim, H., Muhammed, S., & Tarim, M. (2018). Relationship between knowledge management processes and performance: critical role of knowledge utilization in organizations. *Knowledge Management Research & Practice*, 1–15. https://doi.org/10.1080/14778238.2018.1538669.

Assessing the Effectiveness of Mobile Applications in Managing Salt Intake: A Comprehensive Literature Review

Sara Ramdani, Nour El Houda Benkaddour, and Intissar Haddiya

Abstract

Mobile health (mHealth) applications have emerged as essential tools for managing chronic conditions such as hypertension (HTN) by facilitating dietary monitoring and promoting healthier behaviors. This literature review explores the role of mHealth apps in reducing salt intake, a key factor in managing HTN and cardiovascular disease (CVD). While mHealth interventions show potential in raising awareness and promoting behavior change, their efficacy is often limited by short follow-up periods and inadequate control groups. Advances in algorithms of artificial intelligence (AI) offer opportunities for improving the personalization and accuracy of these apps. However, challenges such as data privacy, variability in app-based estimates, and the need for robust clinical trials remain. Future research should focus on long-term studies to assess the sustainability and scalability of these interventions in reducing CVD globally.

Keywords

Mobile health · Chronic diseases · Artificial intelligence · Salt intake

S. Ramdani (✉) · N. El Houda Benkaddour · I. Haddiya
Laboratory of Epidemiology, Clinical Research and Public Health, Faculty of Medicine and Pharmacy of Oujda, Mohammed First University, Oujda, Morocco
e-mail: sara.ramdani001@gmail.com

I. Haddiya
Department of Nephrology, Mohammed VI University Hospital, Faculty of Medicine and Pharmacy of Oujda, Mohammed First University, Oujda, Morocco

1 Introduction

Mobile health (mHealth) applications have increasingly become indispensable tools for enhancing health behaviors, particularly in the areas of dietary management and chronic condition control, such as hypertension (HTN) and diabetes. These applications provide comprehensive support for dietary assessment, physical activity monitoring, and nutrition education, enabling users to make informed decisions about their health (Paramastri et al. 2020). The advent of mHealth technology has transformed the way individuals engage with their health, offering a convenient, accessible, and personalized approach to managing wellness.

In the specific context of HTN, mHealth interventions have shown considerable potential in addressing one of the key modifiable risk factors for cardiovascular disease (CVD): excessive salt intake. While the body of high-quality studies is still developing, existing research indicates that most salt-reduction interventions via mHealth platforms, such as SMS reminders and app-based tracking, are effective in raising awareness about dietary salt consumption and encouraging healthier eating habits (Ali et al. 2019). These interventions aim not only to inform but also to motivate behavioral changes that can lead to significant improvements in blood pressure (BP) control and overall cardiovascular health.

Excessive salt consumption, which starkly contrasts with the dietary patterns of our evolutionary ancestors, poses severe risks to health. Beyond its well-documented role in elevating BP, high salt intake contributes to a range of adverse health outcomes, including endothelial dysfunction, progression of kidney disease, and heightened cardiovascular risk (Batuman 2013). Increasing potassium intake has been identified as an effective countermeasure to these harmful effects, highlighting the importance of balanced electrolyte consumption in maintaining cardiovascular health (Aaron and Sanders 2013).

M. Gouskir et al. (eds.), *Cutting-Edge Artificial Intelligence Applications*, Sustainable Artificial Intelligence-Powered Applications,
https://doi.org/10.1007/978-3-032-00125-2_8

Addressing the global health challenge posed by high salt intake requires a multi-faceted strategy. This includes reducing the consumption of processed foods, which are often high in sodium, and implementing public health initiatives that empower individuals to make informed dietary choices. Public health efforts focused on salt reduction can lead to substantial improvements in population health, particularly by lowering the incidence of CVD, which remain a leading cause of mortality worldwide (Suckling and Swift 2015).

mHealth apps stand out as promising tools in the fight against HTN and CVD by facilitating dietary changes and enhancing awareness of salt intake. However, while the potential of these technologies is clear, further research is essential to establish their long-term efficacy and to refine strategies that maximize their impact on public health. As the field of mHealth continues to grow, so too does the opportunity to make meaningful strides in the prevention and management of chronic diseases through innovative, technology-driven approaches.

2 Methods

We conducted a comprehensive literature search in the PubMed and Web of Science databases without any restrictions on time or geographical scope. The search was performed using the following subject headings: salt intake, hypertension, salt intake applications, and mobile health. To retrieve relevant articles, we applied the search strategy: (salt intake OR hypertension) AND (salt intake applications OR mobile health).

Our review adhered to the 2020 Preferred Reporting Items for Systematic Reviews and Meta-Analyses (PRISMA) guidelines (Fig. 1). Studies that did not focus on essential HTN were excluded from the narrative review. Additionally, articles related to pulmonary, portal, or intracranial hypertension; pre-eclampsia; pregnancy; women; children; infants; adolescents; cancer; and perioperative care were excluded from the study.

3 Overview of Salt Intake and Health

Excessive salt consumption induces various physiological effects that go beyond merely increasing BP and contribute to several organ systems that include the heart, kidneys, vasculature, and brain (Robinson et al. 2019). Increased sodium intake is associated with an enhanced risk of kidney disease and cardiovascular events that are, in general, explained by high levels of inflammation and oxidative stress (Robinson et al. 2019; Malta et al. 2016). High-salt intake contributes to an estimated 5 million deaths worldwide annually. Salt is, however, important for some physiological processes; usually, it also improves the taste of food. However, with the addition of processed food, which contains considerable amounts of salt, control over intake by the individual is diminished. To be effective in decreasing salt intake, the interventions need to be multi-faceted and targeted at all levels of the population. These include government policies, public awareness campaigns, and collaboration with the food industry (Hunter et al. 2022).

Guidelines for salt intake from health organizations have become a bone of contention. While some recommend low sodium intake of below 2.3 g/day for entire populations, others believe there is no solid evidence to support such low levels (Graudal and Jürgens 2018; Mente et al. 2021). Most of the population takes 3–5 g/day of sodium, and this seems to be associated with the lowest risk of CVD and mortality (Mente et al. 2021). Yet, salt intake above 6 g/day is well beyond CVD and mortality and can significantly raise BP (Durack et al. 2008; Legetic and Campbell 2011). By the year 2020, the Pan American Health Organization recommended decreasing salt intake to below 5 g per person per day in the Americas (Legetic and Campbell 2011). These include gradual salt reduction in processed foods and encourage the use of salt replacers, herbs, and spices (Durack et al. 2008). Although debate continues, there exists general consensus that curbing excess salt intake is necessary for public health.

4 Impact on Health Outcomes

Measurement and tracking errors of salt intake are always influenced by elements such as the breadth of food databases, errors in user input, individual variability in metabolic reactions, and limitations of the technology involved. To improve the precision of mobile app salt intake measurement, researchers and developers should place priority on expanding the scope of food databases, developing user interfaces, implementing advanced technologies, as well as carrying out comprehensive validation studies. All of the above would make it possible for mobile apps to become more reliable guides for patients with HTN and other medical conditions in making their dietary selections for an enhancement of the quality of their lives.

There are numerous studies intuitively showing the negative health impacts of high sodium intake, especially on BP and kidney function. Systematic reviews have shown that high salt intake tends to increase BP and heighten the risk for kidney disease (Malta et al. 2016; Johnson et al. 2015). In contrast, sodium reduction lowers BP, especially among hypertensives, and reduces albuminuria in CKD patients (Malta et al. 2016). Smartphone app-based interventions are showing some promise in the management of HTN, and

Fig. 1 Flow chart diagram

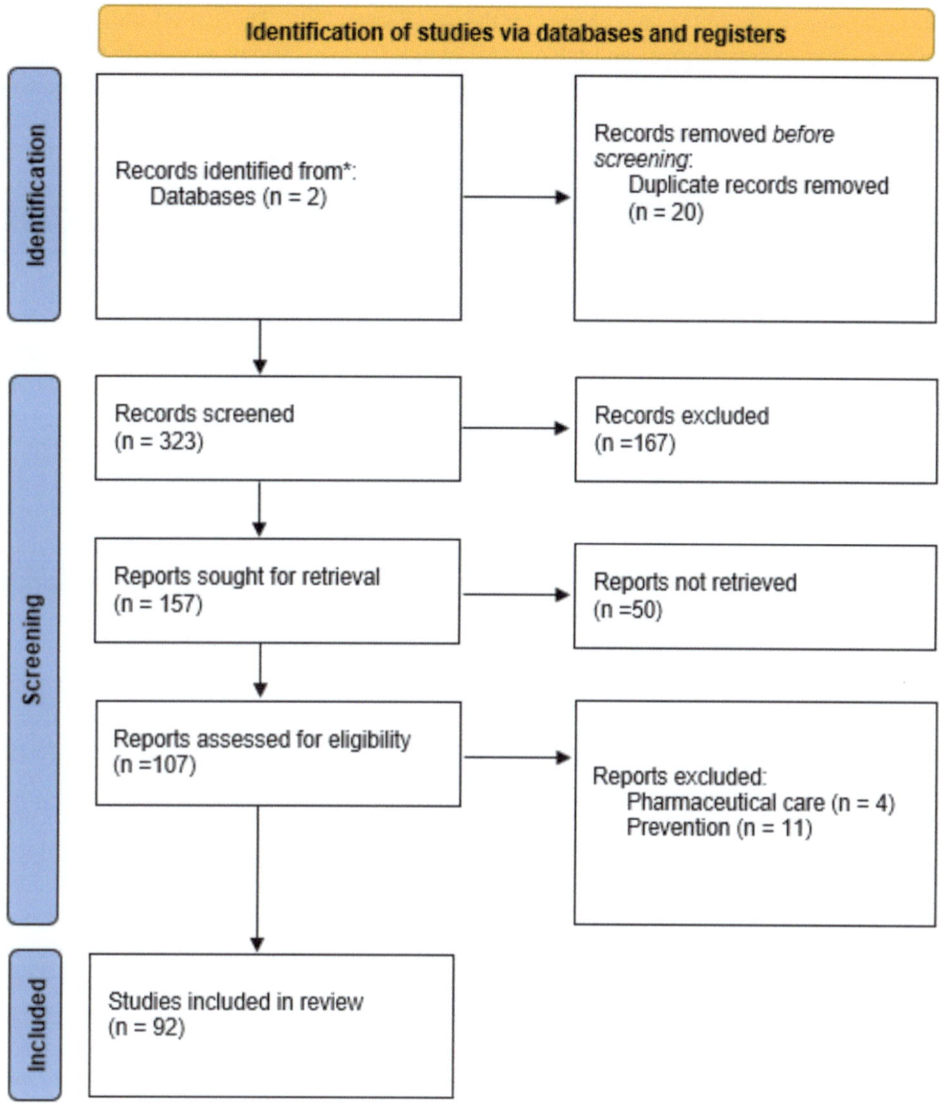

where a meta-analysis of RCTs shows that these interventions lowered systolic PB and improved medication adherence in hypertensive patients (Xu and Long 2020). Yet in this instance, these apps have had no significant effect on levels of physical activity. While there is strong evidence that salt reduction produces better health outcomes, more high-quality studies in this field would be beneficial to the researchers (Malta et al. 2016).

5 Technological Solutions: Phone Applications

Mobile Health (mHealth) interventions hold great promise for CVD prevention worldwide. These include interactive voice response, text messaging, and smartphone applications, which were found to be effective in lifestyle behavior improvement, medication adherence, and disease management (Piette et al. 2015). These approaches showed the potential benefits in areas like weight control, smoking cessation, physical activity, and high BP treatment (Kozik et al. 2021). An integration of propitious wellness approaches and remote check-ups offered by gadgets equipped with artificial intelligence (AI) and advanced health-tracking capabilities (Liaqat et al. 2024). Nonetheless, challenges remain regarding privacy of data, resistance from healthcare providers, and the need for standardization and validation of device data. Future focus should, however, be directed toward optimizing the platform integration for the clinical workflow and concierge issues toward equitable access. While short-term behavior changes are reported, long-term impact on CVD outcomes will require further research (Kelli et al. 2017).

In the last two decades, mHealth applications have so developed, based upon mobile technology, that they can promote healthcare delivery. The field was basically defined by Personal Digital Assistants (PDAs) until 2007, after which it turned toward basic and feature phones for the next 5 years, and thereafter, to smartphones and tablets (Ali et al. 2016). These applications address a variety of healthcare needs, including patient self-management, enhanced communication, health promotion, remote monitoring, and data collection (Fiordelli 2012). The gains expected from mHealth include person-centered health control, generational enhancement of chronic disease supervision vis-a-vis the promotion of healthy behavior. They will also make healthcare easier to access and afford, all the while allowing remote monitoring to minimize in-person visits (Deniz-Garcia et al. 2023). MHealth research has primarily targeted chronic illnesses, yet there is a growing interest in the management of infectious diseases, maternal and child health (Ali et al. 2016). With the progressing mobile technologies, mHealth applications are envisaged not only to identify any weakness and help in extending their healing hand, but also act as a backbone to a sustainable delivery mechanism (Voskarides et al. 2002).

While mobile apps pertaining to physical activity and weight management are emerging in popularity because of their suspected ability to effect behavior change, much research has analyzed features and effectiveness among these apps; most superior differentiated physical activity apps tend to rely on only a few Behavior Change Techniques (BCT) features, and they usually include instruction, modeling, feedback, goal-setting, and social support (Conroy et al. 2014; Yang et al. 2015). The popular weight management apps generally exhibit moderate overall quality, scoring highest for functionalities yet lowest for information quality. Behavioral tracking is generally included in these apps, along with an average of ten BCTs, mainly goal-setting and self-monitoring (Bardus et al. 2016). In a similar manner, salt-reduction apps—SaltSwitch being one such app—provide immediate nutritional information and suggest lower salted alternatives when food barcodes are scanned, realizing an impressive output in salt savings particularly among individuals suffering with CVD (Eyles et al. 2017).

Further research has examined mHealth interventions for salt reduction, identifying health and nutritional education, combined interventions, and salt meters as successful approaches (Silva-Santos et al. 2021). A systematic review found few high-quality studies but pointed out potential effectiveness—if, in fact, it utilizes innovative technologies and large-scale trials with appropriate outcome variables (Ali et al. 2019). For examples, an AppSalt program trialed in China found that the program had consistently high fidelity and retention rates with approximately 80% of participants completing app-based salt-reduction courses (Sun et al. 2022). Moreover, the pilot comparing smartphone apps and traditional methods concluded that the MyFitnessPal users reduced their predicted 24 h sodium excretion much more than those with paper tracking (Ipjian and Johnston 2017).

These results highlight the hope offered by mHealth applications, especially those utilizing technology to strengthen or bolster dietary intake in areas like lower salt consumption. Even though these applications are increasingly popular in mHealth interventions, the quality of these applications still varies widely. Improvements in app quality, utility, and information accuracy are required in order to enhance effectiveness on such issues as these: physical activity, weight control, and dietary changes, again including, but not limited to, salt reduction. Improvements in those aspects will help mHealth technologies better do what they are needed for in public health and traditionally will ensure sustained successful use of mHealth technologies across a gamut of healthcare services (Ali et al. 2016; Fiordelli 2012; Deniz-Garcia et al. 2023; Voskarides et al. 2002).

6 Evaluation of Salt Intake Phone Applications

mHealth apps empower patient self-management and health tracking by enabling them to perform such activities as personalized feedback, data export, and integration with other health apps (Mendiola et al. 2015). These apps exploit smartphone sensors to gather, analyze, and report health-related data, thus providing tailored plans and coaching (Higgins 2016). Although users appreciate simple, time-saving features that are direct instruction-giving and feature data-sharing, such tasks need to be implemented correctly to work (Mendiola et al. 2015). These apps enhance self-awareness and condition self-management by providing customization capabilities and integration into broader health ecosystems, while nonetheless poorly implemented tracking features may deteriorate user experience (Polhemus et al. 2022). Future studies should investigate specific app types, diseases, and behavior changes to further the development and success of the apps (Mendiola et al. 2015; Polhemus et al. 2022).

mHealth apps have great potential for behavior change; however, further delicate evidence is needed on their effectiveness. Studies reveal that commonly incorporated behavior change techniques in apps include self-monitoring, goal-setting, and feedback (Middelweerd et al. 2014; Payne et al. 2015). Personalization, together with prompts and cues, is also an effective component in most interventions (Dugas et al. 2020). Positive outcomes have been noted in most of these apps from small-scale studies but require larger randomized controlled trials to establish their efficacy (Payne et al. 2015; Zhao et al. 2016). Growing user acceptability of health apps makes them a potentially feasible delivery system for interventions (Payne et al. 2015). Yet the quality of available apps varies greatly, with many of them unable to take advantage of well-established theories of behavior change and advanced mobile technologies (Middelweerd et al. 2014;

Dugas et al. 2020). Future research should then involve identifying the most effective features of an app and the behavior change strategy used to facilitate stubborn health behavior change.

mHealth apps have become useful in the area of dietary behavior change: these applications integrate tracking of physical activities and some renowned nutrition apps making use of food diaries with textual search and a barcode scanner for data entry (Franco et al. 2016). Such apps have been shown to be effective in bringing about changes in diet, with studies showing that app users had larger sodium intake reductions than paper trackers did (Ipjian and Johnston 2017). Using smartphone tools like a global positioning system and accelerometers, these apps offer personalized coaching and allow users to show off their accomplishments on social media (Higgins 2016). At present, however, the focus for these apps seems to revolve largely around energy balance and nutrient intake, while emerging technologies like image recognition and AI open avenues for more individualized dietary recommendations (Franco et al. 2016).

Novel research has concerned with several methods to decreasing salt intake, using technology-based interventions. The food search interfaces began with abstract and simple visual or textual explanations, which affected salty food preferences, while an avatar-based intervention of future self-proved promising to increasing the awareness of risks and encouraging low-sodium dieting (Berge et al. 2108; Fuchs et al. 2019). The smartphone app-based education program was effective in reducing salt intake in schools of China, which lowered systolic BP in adults but had no significant effects in children (He et al. 2022). There were proposals for PDA applications using a user-centered design approach meant for dialysis patients, having strict sodium restrictions which support interfaces without stylus input, writing, or small icons, keeping in mind various levels of visual acuity, literacy skills, and computer experience (Connelly et al. 2005). These findings bring to sharp relief the efficacy of technology in helping a wide variety of populations stick to low-sodium diets.

Additionally, studies have evaluated the effectiveness of smartphone applications in measuring and tracking dietary sodium intake. Jung et al. found a strong correlation between sodium intakes measured using a smartphone app and 24 h urine sodium levels (Jung et al. 2023). However, Griffiths et al. reported that nutrient calculations from popular nutrition tracking apps are often lower than those obtained from research-grade software (Griffiths et al. 2018). Ipjian and Johnston demonstrated that using a smartphone app led to greater reductions in predicted 24 h sodium excretion compared to paper-based tracking, with users reporting higher satisfaction with the app (Ipjian and Johnston 2017). However, Parkington and Roussos highlighted the need for more precise definitions of salt consumption and more sensitive measurement tools to improve the validity of salt-eating behavior assessments (Parkington and Roussos 2008). While these studies suggest that smartphone apps have promise in facilitating dietary sodium reduction, further research is needed to enhance their accuracy and effectiveness in practical settings.

7 Future Directions

Recent research underscores the growing potential of AI-powered smartphone applications in promoting healthier dietary behaviors, particularly in reducing salt intake. One study, for instance, demonstrated the effectiveness of an AI-based app in significantly decreasing salt consumption among CKD patients over a 3-month period (Yanai et al. 2023). This finding is particularly noteworthy as managing salt intake is critical for CKD patients, who are at heightened risk for HTN and cardiovascular complications. The app's success highlights the potential for AI-driven tools to offer tailored interventions that align with individual health needs, thereby improving patient outcomes.

Recent studies have evaluated AI algorithms for the management of individualized salt intake. Ryu et al. have developed a hybrid CNN-based model with YOLOv4 and ResNet-101 for sodium intake estimation from food images for hospitalized patients (Ryu et al. 2024). Yanai et al. have reviewed an AI-based smartphone application showing effective reduction of short-term salt intake in the case of chronic kidney disease patients (Yanai et al. 2023). Guzman-Vilca et al. devised a machine learning (ML) model to predict population-level salt consumption in low- and middle-income countries using age, sex, and BP as the simple predictors (Guzman-Vilca et al. 2021). Lee et al. have proposed an AI-based precise nutrient analysis model assessing 24 nutrients based on names and serving sizes of dishes and using semantic and nutritional analysis models with cosine similarity algorithms (Lee et al. 2022).

How these studies prove the potentiality of AI algorithms by providing better personalization and accuracy in salt intake estimation and management across different healthcare settings and populations remains a mystery. Recent studies thus underscore the ever-growing potential of AI-enabled smartphone applications in mitigating poor dietary behavior through salt reduction. For example, in another study, the effectiveness of the AI-based application occurred in the significant reduction of salt consumption among CKD patients over a 3-month period (Payne et al. 2015). The observation is critical in the sense that the control of salt intake is essential to CKD patients who are at high risk for HTN and cardiac complications. The good results of the application point to the potential for AI technologies to provide individualized interventions to meet personal health needs and improve patient outcomes.

Further advancements in AI and mHealth technologies are reflected in the development of comprehensive app-based interventions. This initiative, building on previous successful interventions, proposes a multi-faceted approach to long-term salt reduction, particularly targeting schoolchildren and their families. The program incorporates elements of education, target setting, monitoring, and decision support, all designed to encourage sustained behavioral change (He et al. 2019). The use of AI in this context allows for personalized recommendations and feedback, making it easier for users to adhere to healthier eating habits over time.

Recent studies have explored various AI algorithms for managing salt intake and related applications. Fuzzy inference systems (FIS) and adaptive network-based fuzzy inference systems (ANFIS) have shown promise in predicting saltwater intrusion in coastal aquifers, outperforming genetic programming models (Roy and Datta 2018). In hospital settings, a hybrid AI model combining YOLOv4, ResNet-101, and hyperspectral imaging has been used to estimate patients' sodium intake from meal images, correlating with 24 h urine sodium measurements (Ryu et al. 2024). Deep learning approaches have also been applied to uncover hidden salt effects (Manolis and Kallistratos 2023). In irrigation management, artificial neural networks (ANNs) and fuzzy logic-based systems have been widely used, with ANNs being particularly popular for ML applications. Hybrid neuro-fuzzy approaches combine the strengths of both techniques, while wireless sensor networks and computer vision with remote sensing images show promise for future developments (Alvim et al. 2022).

Alongside those advancements, however, nutrition-related applications are utilizing ML techniques. For example, ML models have been developed to estimate salt intake at the population level using simple predictors such as age, sex, and BP, serving as a cost-effective alternative to urine sample analysis, which is more invasive and resource-intensive (Guzman-Vilca et al. 2021, 2022). These models thus help justify public health efforts to reduce dietary salt intake, guiding public policy-directed initiatives based on the estimation of daily correspondence from most effective to least effective methods to track sodium intake in children. On the other hand, ML models performed well in predicting 24 h urinary sodium excretion from self-reported questionnaire data, which can improve the accuracy of sodium assessments limited by self-reported methods (Hamaya et al. 2024). In pediatric nutrition, ML technology is being utilized to develop smartphone apps that allow the tracking of dietary behavior with appropriate healthcare advice at real time, thus enabling intervention at an early stage of malnutrition and the risks of obesity (Young et al. 2024). And, in the future, as these technologies continue to mature, they have the potential to transform nutritional science and public health.

More and more various smartphone applications that have been developed are aimed at providing advice to the heart disease and chronic kidney disease patients to start taking low-salt diets. These apps adopt AI-driven dietary assessments, nutritional barcode scanning, and theories on behavior change (Yanai et al. 2023; Perin et al. 2022). Research shows such app-based interventions result in clinically significant short-term reductions in salt intake, considerably more than traditional methods (Ipjian and Johnston 2017; Yanai et al. 2023). Compared to the seemingly traditional paper-based techniques, users showed higher satisfaction with app-based monitoring systems, citing an evident convenience in the immediate feedback of the technology (Ipjian and Johnston 2017). In addition to augmenting management of salt intake, these apps could possibly help enhance self-care practices and open up professional relationships with their patients in a more informed and engaged healthcare interaction (Perin et al. 2022).

Despite the encouraging initial results, there is an urgent need for more research to be carried out in order to judge the long-term success and scalability of these interventions. Although current app-based solutions are effective for short-term management, several studies argue for rigorous clinical trials to investigate the possibility of dietary adaptations over the long-term and enhanced overall health (Yanai et al. 2023; Eyles et al. 2014). Mixed results have been reported in relation to user engagement and overall effectiveness in long-term studies of mHealth applications. In general, although mHealth interventions can strengthen short-term physical activity and also maintain effect over the long run, the absolute magnitude of the effect is likely to reduce with time (Mönninghoff et al. 2021). Continued app use is reliant on user experience, whereby good experiences are noted to usually happen in the first few weeks of usage (Biduski et al. 2020). What remains ambiguous is why long-term engagement with an application is so problematic, where it could be in response to technical difficulties, usability issues, or lack of support, among other issues (Locatelli et al. 2024). Recent studies have looked at mHealth-based interventions for salt reduction in the management of HTN. A feasibility randomized controlled trial showed the acceptability and feasibility of a mobile app-based intervention for salt reduction in primary care settings (Riches et al. 2021). The intervention, including individual advising and the smartphone app, was encouraging in its ability to affect salt behaviors. A systematic review identified the need for more robust, large-scale trials with proper outcome measures, particularly emphasizing on 24 h urine sodium, in order to assess the effect of mHealth interventions for salt reduction (Ali et al. 2019). Another ongoing randomized controlled trial SaltSwitch wishes to study the efficiency of a smartphone app catering to lower salt food choices for those suffering from CVD(Eyles et al. 2014). Taken together, these studies show the evolution of trial design and the testing of the efficacy of mHealth apps in managing salt intake.

User assessment of the app's capabilities and persistence in health goals are key dimensions influencing continued use decisions (Vaghefi and Tulu 2019). To improve long-term engagement, mHealth app designers should consider user characteristics, motivations, and technological abilities, as well as incorporate features like visual resources and practicality in treatment monitoring (Biduski et al. 2020; Locatelli et al. 2024).

Looking ahead, future research on dietary salt intake should prioritize longitudinal studies and large-scale clinical trials to address the existing gaps in knowledge. There is a critical need for interventional studies that can clarify the relationship between sodium intake and cardiovascular risk, particularly in light of the confounding factors present in current observational data (Zoccali and Mallamaci 2016). In the same way, a lot of important studies showing no association between sodium intake and CVD are on non-diabetics. Current guidelines on sodium intake and medications often come from studies on non-diabetic individuals, after long-term follow-up (Horikawa and Sone 2017). Ha refers to the residual confounding potential risk and states future studies should explore the relationship between quantity of salt and cardiovascular health (Ha 2014). According to Overwyk et al., it would be most useful to standardize the methodologies of measuring sodium intake since differences in the study designs have hampered the field's understanding of sodium's health effects (Overwyk et al. 2020).

In summary, the development of AI algorithms into mHealth applications holds significant promise for advancing public health, particularly in the context of salt intake reduction. However, realizing this potential requires a concerted effort to conduct rigorous research that can inform evidence-based guidelines and interventions. By addressing the current gaps in knowledge and refining these digital tools, we can better equip individuals to make healthier dietary choices and, ultimately, reduce the global burden of CVD.

8 Challenges and Limitations

Current research on privacy in relation to health apps and user data privacy on salt intake raises significant concerns, and as such, there is a fine line to walk between harnessing technology for health and ensuring the privacy of personal data related to user health. Moreover, while these health apps have been and are successfully enhancing healthy behavior (e.g., reducing salt intake which is typically associated with more positive health outcomes), users have retained a great concern for the privacy of their information (Fuchs et al. 2019). In particular, mature adults seem to worry that their private health information rather than personal identifiable information would be usurped for the advertisement purposes (Yanai et al. 2023). Thus, while users understand there is the possibility of misuse related to their data privacy, they are primarily concerned with the potentially problematic data use with them being more disturbed with ads. And, while worrying about the privacy of their personal data, it appears that developers are faced with a unique challenge in regard to the use of privacy making sure user agreements are in compliance, continually weighing this dilemma of effective privacy policies and advertising with standard practice (Balebako et al. 2014). However, despite these are concerns, app data collection and usage offer possible advantages to provide academically viable reach, more accurate randomization, and improved data collection procedures in research (Zhang et al. 2018).

Although user do express a desire to prudently protect their health-related data, they equally sarcastically describe low levels of comfort or annoyance toward the data protection data practice (Ryu et al. 2024). This annoyance of increased comfortability in using health over fitness apps seems to create an intent to engage in further behavior that creates irrational conflict, while considering the discomfort in acceptance or using this technology, users are more likely considering the dangers regarding accept data and irrevocable misuse (Yanai et al. 2023). Experts advise the users not to presume that whatever data is being shared with health and fitness applications is automatically private and protected (Guzman-Vilca et al. 2021). The gap between users' worries and behavior exemplifies the most urgent necessity for more user-friendly and effective ways of data protection. Furthermore, enhancing understanding of the dangers associated with sharing sensitive health information through mobile apps is of great importance so that users might be better aware of the consequences.

Developing mobile apps comes with significant challenges, particularly when it comes to handling user data responsibly and complying with privacy regulations. Research has shown that both Android and iOS apps often fall short of their own stated privacy policies, potentially putting users' sensitive information at risk (Kununka et al. 2017). Smaller app development companies are less likely to implement robust privacy and security practices, and many developers lack awareness of data collection by third-party tools (Balebako et al. 2014). Analyzing popular Android apps has revealed widespread data-sharing across multiple domains, often without adequately informing users. To tackle these issues, developers are encouraged to embrace privacy-by-design principles, adopt opt-in data-sharing models, and ensure transparency about how user data is managed (Brandtzaeg et al. 2019). Additionally, a technical guideline catalog has been proposed to help developers bridge the gap between privacy laws and real-world implementation, ensuring compliance with regulations such as the EU GDPR (Hatamian 2020).

Beyond privacy concerns, research has also identified major gaps in the effectiveness of mHealth interventions aimed at reducing salt intake. While mHealth technologies have the potential to help people cut down on salt consumption, many of these interventions suffer from small sample sizes, lack of control groups, and short follow-up periods. A systematic review of mHealth strategies for salt reduction identified 11 studies, but most lacked the rigor needed to draw reliable conclusions about their effectiveness. Current interventions tend to focus on raising awareness of dietary salt intake, often relying on SMS-based platforms to deliver health information (Ali et al. 2019). However, general consumer awareness about recommended daily salt intake remains low, and many people are unaware of the primary dietary sources of salt (Lee et al. 2022).

To close these knowledge gaps, validated tools are needed to assess how salt awareness impacts dietary behavior (Lee et al. 2022). A feasibility randomized controlled trial conducted among hypertensive patients showed that a mobile app designed for salt reduction was well received and met the necessary criteria to proceed to a larger trial. However, the study was not sufficiently powered to detect significant changes in salt intake or BP (Riches et al. 2021). Meanwhile, ongoing research—including a randomized controlled trial in Brazil—is assessing the effectiveness of mobile app interventions specifically designed to promote healthy salt consumption (Fuchs et al. 2019). Addressing privacy concerns and improving intervention effectiveness are crucial to ensuring the long-term success of these digital health solutions, especially since researchers have yet to fully explore the potential of mHealth apps beyond simple treatment interventions (Zhang et al. 2018).

ML has the potential to revolutionize both nutritional epidemiology and personalized dietary recommendations. By modeling complex diagnostic criteria and extracting meaningful information from unstructured electronic health records, ML could significantly improve how health outcomes are identified (Wong et al. 2018). In the field of nutrition science, ML can help overcome challenges such as measurement errors, dietary complexity, and confounding factors by enabling larger sample analyses, automating food classification, and modeling nonlinear dietary relationships (Morgenstern et al. 2021). More specifically, ML techniques can improve our understanding of dietary patterns and their connections to health outcomes, potentially leading to the development of stronger, evidence-based dietary guidelines (Bodnar et al. 2022). Personalized dietary recommendations powered by ML have already demonstrated better outcomes in lipid profile improvement, blood glucose control, and weight management compared to conventional methods (Varshney et al. 2023). However, challenges such as data quality issues, algorithmic biases, and ethical concerns within the healthcare system still need to be addressed. Future research should focus on longitudinal studies, real-time data collection, and the psychological factors influencing diet to advance the field.

The potential of ML in nutrition research extends beyond individual recommendations. ML algorithms can enhance dietary measurement accuracy by automatically classifying food images and analyzing complex dietary patterns (Morgenstern et al. 2021). They also help address challenges in interpreting nutritional data, such as capturing food interactions and nonlinear relationships (Bodnar et al. 2022). A practical example of this is the study by Guzman-Vilca et al., who developed an ML model capable of estimating population-level salt intake using simple predictors like age, sex, and BP. When applied to 54 countries, this model showed promising accuracy, filling a crucial gap in global salt consumption data. The study found that Western Pacific countries had the highest predicted salt intake, whereas African countries had the lowest (Guzman-Vilca et al. 2022). These advancements in ML have the potential to transform dietary tracking and inform evidence-based nutritional policies.

Despite the promise of mHealth interventions, several limitations can affect the validity of findings. Many studies lack adequate control groups and have short follow-up periods, making it difficult to assess the long-term impact and scalability of these interventions (Marcolino et al. 2018). Acknowledging these limitations is essential for scientific progress, as it allows researchers to refine methodologies and avoid misinterpretations by readers and reviewers (Puhan et al. 2012). Improving the quality of mHealth research will require stronger study designs, better control groups, and individualized patient data analyses to ensure more accurate and meaningful results (Chen et al. 2012).

In conclusion, while mHealth apps offer great potential for promoting healthier behaviors, including salt reduction and chronic disease management, there are still major hurdles to overcome. Challenges related to data privacy, consumer knowledge gaps, and the overall effectiveness of interventions must be addressed. A collaborative effort is needed to improve user data protection, educate the public, and conduct high-quality research that validates the efficacy of these digital health tools. Future interventions should leverage emerging technologies, prioritize preventative care, and conduct large-scale, rigorous clinical trials using robust outcome measures, such as 24 h urine sodium levels, to maximize the potential of mHealth in public health (Ali et al. 2019; Byrd-Bredbenner et al. 2017; Johnson et al. 2017; Shelton et al. 2018).

9 Conclusion

In conclusion, mobile applications have become indispensable tools for individuals managing HTN and other health conditions, particularly in the context of reducing sodium intake. These apps, which track dietary habits and provide real-time feedback, empower users to make informed decisions, thereby supporting effective salt-reduction initiatives crucial for mitigating HTN and CVD risks. However, the accuracy of salt intake measurement and tracking in mobile apps remains a critical area of ongoing research, as variability in app-based estimates compared to traditional methods highlights challenges such as user input errors, the accuracy of food databases, and individual differences in dietary habits. To enhance the reliability of these tools, it is essential for researchers and developers to focus on expanding and updating food databases, improving user interfaces, leveraging advanced technologies, and conducting thorough validation studies. Moreover, the integration of mobile apps with broader strategies, such as engaging the food industry in reformulating processed foods, implementing legislative measures, and ensuring iodine fortification, will be essential in achieving effective sodium reduction and improving population health outcomes.

References

Aaron KJ, Sanders PW. Role of dietary salt and potassium intake in cardiovascular health and disease: a review of the evidence. Mayo Clinic proceedings. (2013);88(9):987–95. https://doi.org/10.1016/j.mayocp.2013.06.005

Ali SH, Luo R, Li Y, Liu X, Tang C, Zhang P. Application of Mobile Health Technologies Aimed at Salt Reduction: Systematic Review. JMIR mHealth and uHealth. (2019);7(4):e13250. https://doi.org/10.2196/13250

Ali EE, Chew L, Yap KY-L. Evolution and current status of mhealth research: a systematic review. BMJ Innovations. (2016);2(1).

Alvim SJ, Guimarães CM, Sousa EFd, Garcia RF, Marciano CR. Application of artificial intelligence for irrigation management: a systematic review. Engenharia Agrícola. (2022);42:e20210159.

Balebako R, Marsh A, Lin J, Hong J, Cranor L. The Privacy and Security Behaviors of Smartphone App Developers 2014.

Bardus M, van Beurden SB, Smith JR, Abraham C. A review and content analysis of engagement, functionality, aesthetics, information quality, and change techniques in the most popular commercial apps for weight management. International Journal of Behavioral Nutrition and Physical Activity. (2016);13:1–9.

Batuman V. Salt and hypertension: why is there still a debate? Kidney international supplements. (2013);3(4):316–20. https://doi.org/10.1038/kisup.2013.66

Berge A, Sjøen VV, Starke AD, Trattner C. Changing salty food preferences with visual and textual explanations in a search interface. arXiv preprint arXiv:2108.01427. (2021).

Biduski D, Bellei EA, Rodriguez JPM, Zaina LAM, De Marchi ACB. Assessing long-term user experience on a mobile health application through an in-app embedded conversation-based questionnaire. Computers in Human Behavior. (2020);104:106169.

Bodnar LM, Kirkpatrick SI, Naimi AI. Machine learning can improve the development of evidence-based dietary guidelines. Public health nutrition. (2022);25(9):2566–9.

Brandtzaeg PB, Pultier A, Moen GM. Losing control to data-hungry apps: A mixed-methods approach to mobile app privacy. Social Science Computer Review. (2019);37(4):466–88.

Byrd-Bredbenner C, Wu F, Spaccarotella K, Quick V, Martin-Biggers J, Zhang Y. Systematic review of control groups in nutrition education intervention research. International Journal of Behavioral Nutrition and Physical Activity. (2017);14:1–26.

Chen C, Haddad D, Selsky J, et al. Making sense of mobile health data: an open architecture to improve individual-and population-level health. Journal of medical Internet research. (2012);14(4):e2152.

Connelly KH, Rogers Y, Siek KA, et al., editors. Designing a PDA interface for dialysis patients to monitor diet in their everyday life. Proc. of HCI-International; 2005.

Conroy DE, Yang C-H, Maher JP. Behavior change techniques in top-ranked mobile apps for physical activity. American journal of preventive medicine. (2014);46(6):649–52.

Deniz-Garcia A, Fabelo H, Rodriguez-Almeida AJ, et al. Quality, Usability, and Effectiveness of mHealth Apps and the Role of Artificial Intelligence: Current Scenario and Challenges. Journal of medical Internet research. (2023);25:e44030. https://doi.org/10.2196/44030

Dugas M, Gao G, Agarwal R. Unpacking mHealth interventions: a systematic review of behavior change techniques used in randomized controlled trials assessing mHealth effectiveness. Digital health. (2020);6:2055207620905411.

Durack E, Alonso-Gomez M, Wilkinson MG. Salt: a review of its role in food science and public health. Current Nutrition & Food Science. (2008);4(4):290–7.

Eyles H, McLean R, Neal B, Doughty RN, Jiang Y, Mhurchu CN. Using mobile technology to support lower-salt food choices for people with cardiovascular disease: protocol for the SaltSwitch randomized controlled trial. BMC public health. (2014);14:1–8.

Eyles H, McLean R, Neal B, et al. A salt-reduction smartphone app supports lower-salt food purchases for people with cardiovascular disease: findings from the SaltSwitch randomised controlled trial. European journal of preventive cardiology. (2017);24(13):1435–44.

Fiordelli M, editor Mapping M-Health: a Decade of Evolution. Medicine 2.0 Conference; 2012: JMIR Publications Inc., Toronto, Canada.

Franco RZ, Fallaize R, Lovegrove JA, Hwang F. Popular nutrition-related mobile apps: a feature assessment. JMIR mHealth and uHealth. (2016);4(3):e5846,

Fuchs K, Meusburger D, Haldimann M, Ilic A. NutritionAvatar: designing a future-self avatar for promotion of balanced, low-sodium diet intention: framework design and user study. Proceedings of the 13th Biannual Conference of the Italian SIGCHI Chapter: Designing the next interaction; Padova, Italy: Association for Computing Machinery; 2019. p. Article 11.

Graudal N, Jürgens G. Conflicting evidence on health effects associated with salt reduction calls for a redesign of the salt dietary guidelines. Progress in Cardiovascular Diseases. (2018);61(1):20–6.

Griffiths C, Harnack L, Pereira MA. Assessment of the accuracy of nutrient calculations of five popular nutrition tracking applications. Public health nutrition. (2018);21(8):1495–502

Guzman-Vilca WC, Castillo-Cara M, Carrillo-Larco RM. Development, validation, and application of a machine learning model to estimate salt consumption in 54 countries. Elife. (2022);11:e72930.

Guzman-Vilca WC, Castillo-Cara M, Carrillo-Larco RM. Estimating salt consumption in 49 low-and middle-income countries: Development, validation and application of a machine learning model. medRxiv. (2021):2021.08. 31.21262944.

Ha SK. Dietary salt intake and hypertension. Electrolytes & Blood Pressure: E & BP. (2014);12(1):7.

Hamaya R, Wang M, Juraschek SP, et al. Prediction of 24-Hour Urinary Sodium Excretion Using Machine-Learning Algorithms. Journal of the American Heart Association. (2024);13(10):e034310.

Hatamian M. Engineering privacy in smartphone apps: A technical guideline catalog for app developers. IEEE Access. (2020);8:35429–45.

He FJ, Zhang P, Luo R, et al. An Application-based programme to reinforce and maintain lower salt intake (AppSalt) in schoolchildren and their families in China. BMJ open. (2019);9(7):e027793.

He FJ, Zhang P, Luo R, et al. App based education programme to reduce salt intake (AppSalt) in schoolchildren and their families in China: parallel, cluster randomised controlled trial. BMJ (Clinical research ed.). (2022);376:e066982. https://doi.org/10.1136/bmj-2021-066982

Higgins JP. Smartphone applications for patients' health and fitness. The American journal of medicine. (2016);129(1):11–9.

Horikawa C, Sone H. Dietary salt intake and diabetes complications in patients with diabetes: An overview. Journal of general and family medicine. (2017);18(1):16–20.

Hunter RW, Dhaun N, Bailey MA. The impact of excessive salt intake on human health. Nature Reviews Nephrology. (2022);18(5):321–35.

Ipjian ML, Johnston CS. Smartphone technology facilitates dietary change in healthy adults. Nutrition. (2017);33:343–7.

Johnson C, Raj TS, Trudeau L, et al. The science of salt: a systematic review of clinical salt studies 2013 to 2014. The Journal of Clinical Hypertension. (2015);17(5):401–11.

Johnson C, Santos JA, McKenzie B, et al. The science of salt: a regularly updated systematic review of the implementation of salt reduction interventions (September 2016–February 2017). The Journal of Clinical Hypertension. (2017);19(10):928–38.

Jung C-Y, Kim Y, Kim HW, et al. Effectiveness of a Smartphone Application for Dietary Sodium Intake Measurement. Nutrients. (2023);15(16):3590.

Kelli HM, Witbrodt B, Shah A. The future of mobile health applications and devices in cardiovascular health. European medical journal. Innovations. (2017);2017:92.

Kozik M, Isakadze N, Martin SS. Mobile health in preventive cardiology: current status and future perspective. Current opinion in cardiology. (2021);36(5):580–8.

Kununka S, Mehandjiev N, Sampaio P. A comparative study of Android and iOS mobile applications' data handling practices versus compliance to privacy policy. Privacy and Identity Management. The Smart Revolution: 12th IFIP WG 9.2, 9.5, 9.6/11.7, 11.6/SIG 9.2. 2 International Summer School, Ispra, Italy, September 4–8, 2017, Revised Selected Papers 12. (2018):301–13.

Lee H-A, Huang T-T, Yen L-H, et al. Precision nutrient management using artificial intelligence based on digital data collection framework. Applied Sciences. (2022);12(9):4167.

Legetic B, Campbell N. Reducing salt intake in the Americas: Pan American health organization actions. Journal of health communication. (2011);16(sup2):37–48.

Liaqat M, Mushtaq M, Jamil A, et al. Mobile Health Interventions: A Frontier for Mitigating the Global Burden of Cardiovascular Disease. Cureus. (2024);16(6).

Locatelli JK, Scortegagna SA, Marchi ACBD. The long-term user experience of an m-Health application. (2024).

Malta D, Petersen KS, Johnson C, et al. High sodium intake increases blood pressure and risk of kidney disease. From the Science of Salt: A regularly updated systematic review of salt and health outcomes (August 2016 to March 2017). The Journal of Clinical Hypertension. (2018);20(12):1654–65.

Manolis AJ, Kallistratos MS. Deep learning approach to unmask hidden salt effects in the era of artificial intelligence. Oxford University Press US; 2023. p. 4458–60.

Marcolino MS, Oliveira JAQ, D'Agostino M, Ribeiro AL, Alkmim MBM, Novillo-Ortiz D. The impact of mHealth interventions: systematic review of systematic reviews. JMIR mHealth and uHealth. (2018);6(1):e8873.

Mendiola MF, Kalnicki M, Lindenauer S. Valuable features in mobile health apps for patients and consumers: content analysis of apps and user ratings. JMIR mHealth and uHealth. (2015);3(2):e4283.

Mente A, O'Donnell M, Yusuf S. Sodium intake and health: what should we recommend based on the current evidence? Nutrients. (2021);13(9):3232.

Middelweerd A, Mollee JS, van der Wal CN, Brug J, Te Velde SJ. Apps to promote physical activity among adults: a review and content analysis. International journal of behavioral nutrition and physical activity. (2014);11:1–9.

Mönninghoff A, Kramer JN, Hess AJ, et al. Long-term effectiveness of mHealth physical activity interventions: systematic review and meta-analysis of randomized controlled trials. Journal of medical Internet research. (2021);23(4):e26699.

Morgenstern JD, Rosella LC, Costa AP, de Souza RJ, Anderson LN. Perspective: Big data and machine learning could help advance nutritional epidemiology. Advances in Nutrition. (2021);12(3):621–31.

Overwyk KJ, Quader ZS, Maalouf J, et al. Dietary sodium intake and health indicators: a systematic review of published literature between January 2015 and December 2019. Advances in Nutrition. (2020);11(5):1174–200.

Paramastri R, Pratama SA, Ho DKN, et al. Use of mobile applications to improve nutrition behaviour: A systematic review. Computer methods and programs in biomedicine. (2020);192:105459. https://doi.org/10.1016/j.cmpb.2020.105459

Parkington SR, Roussos S. Measuring salt consumption to guide behavior change in applied settings: a critical review. American Journal of Health Promotion. (2008);23(2):101–7.

Payne HE, Lister C, West JH, Bernhardt JM. Behavioral functionality of mobile apps in health interventions: a systematic review of the literature. JMIR mHealth and uHealth. (2015);3(1):e3335.

Perin MS, São-João T, Gallani MCBJ, Agbadje TT, Rodrigues RCM, Cornélio ME. A Mobile phone app intervention to promote healthy salt intake among adults: protocol for a randomized controlled study. JMIR research protocols. (2022);11(6):e37853.

Piette JD, List J, Rana GK, Townsend W, Striplin D, Heisler M. Mobile health devices as tools for worldwide cardiovascular risk reduction and disease management. Circulation. (2015);132(21):2012–27.

Polhemus A, Simblett S, Dawe-Lane E, et al. Health tracking via mobile apps for depression self-management: qualitative content analysis of user reviews. JMIR Human Factors. (2022);9(4):e40133.

Puhan MA, Akl EA, Bryant D, Xie F, Apolone G, ter Riet G. Discussing study limitations in reports of biomedical studies—the need for more transparency. Health and quality of life outcomes. (2012);10:23. https://doi.org/10.1186/1477-7525-10-23

Riches SP, Piernas C, Aveyard P, et al. A mobile health salt reduction intervention for people with hypertension: results of a feasibility randomized controlled trial. JMIR mHealth and uHealth. (2021);9(10):e26233.

Robinson AT, Edwards DG, Farquhar WB. The influence of dietary salt beyond blood pressure. Current hypertension reports. (2019);21:1–11.

Roy DK, Datta B. Comparative efficiency of different artificial intelligence based models for predicting density dependent saltwater intrusion processes in coastal aquifers and saltwater intrusion management utilizing the best performing model. Desalination and Water Treatment. (2018);105(2018):160–80.

Ryu J, Kim S, Lim Y, et al. Sodium Intake Estimation in Hospital Patients Using AI-Based Imaging: Prospective Pilot Study. JMIR Formative Research. (2024);8(1):e48690.

Shelton RC, Cooper BR, Stirman SW. The sustainability of evidence-based interventions and practices in public health and health care. Annual review of public health. (2018);39(1):55–76.

Silva-Santos T, Moreira P, Rodrigues M, et al. Interventions that successfully reduced adults salt intake—a systematic review. Nutrients. (2021);14(1):6.

Suckling RJ, Swift PA. The health impacts of dietary sodium and a low-salt diet. Clinical medicine. (2015);15(6):585–8.

Sun Y, Li Y, He FJ, et al. Process evaluation of an application-based salt reduction intervention in school children and their families (Appsalt) in China: a mixed-methods study. Frontiers in Public Health. (2022);10:744881.

Vaghefi I, Tulu B. The continued use of mobile health apps: insights from a longitudinal study. JMIR mHealth and uHealth. (2019);7(8):e12983.

Varshney N, Jadhav N, Gupta K, Mate NR, Rose A, Kumar P, editors. Personalized Dietary Recommendations Using Machine Learning: A Comprehensive Review. 2023 International Conference on Artificial Intelligence for Innovations in Healthcare Industries (ICAIIHI); 2023: IEEE.

Voskarides S, Pattichis CS, Istepanian RSH, Kyriacou E, Pattichis MS, Schizas CN. Mobile health systems: A brief overview. Digital Wireless Communications IV. (2002);4740:124–31.

Wong J, Murray Horwitz M, Zhou L, Toh S. Using machine learning to identify health outcomes from electronic health record data. Current epidemiology reports. (2018);5:331–42.

Xu H, Long H. The effect of smartphone app–based interventions for patients with hypertension: systematic review and meta-analysis. JMIR mHealth and uHealth. (2020);8(10):e21759.

Yanai A, Uchiyama K, Suganuma S. Salt Reduction Using a Smartphone Application Based on an Artificial Intelligence System for Dietary Assessment in Patients with Chronic Kidney Disease: A Single-Center Retrospective Cohort Study. Kidney and Dialysis. (2023);3(1):139–51.

Yang C-H, Maher JP, Conroy DE. Implementation of behavior change techniques in mobile applications for physical activity. American journal of preventive medicine. (2015);48(4):452–5.

Young A, Johnson MJ, Beattie RM. The use of machine learning in paediatric nutrition. Current Opinion in Clinical Nutrition & Metabolic Care. (2024);27(3):290–6.

Zhang J, Calabrese C, Ding J, Liu M, Zhang B. Advantages and challenges in using mobile apps for field experiments: A systematic review and a case study. Mobile Media & Communication. (2018);6(2):179–96.

Zhao J, Freeman B, Li M. Can mobile phone apps influence people's health behavior change? An evidence review. Journal of medical Internet research. (2016);18(11):e287.

Zoccali C, Mallamaci F. Moderator's view: Salt, cardiovascular risk, observational research and recommendations for clinical practice. Nephrology Dialysis Transplantation. (2016);31(9):1405–8.

Developments in Smart City Traffic Monitoring Algorithms with a Focus on GC-YOLOv9

Lahcen Gouskir, Mohamed Baslam, Mohamed Gouskir, Abdelmoula Abouhilal, Soufiane Belhouideg, and Hanaa Hachimi

Abstract

In modern urban environments, efficient traffic management is a critical component of smart cities. The increasing complexity of traffic systems necessitates advancements in real-time monitoring technologies. Among these, the YOLO series of algorithms, specifically the GC-YOLOv9, stands out for its enhanced performance in object detection. This article delves into the innovations in traffic monitoring algorithms, particularly focusing on the GC-YOLOv9 model. The GC-YOLOv9 introduces ghost convolution technology, significantly improving the model's accuracy and efficiency. Additionally, this paper discusses the integration of IoT frameworks with edge computing to enable real-time data processing, further enhancing smart city infrastructure.

Keywords

Smart cities · Traffic monitoring · YOLOv9 · Internet of things · Real-time data processing

L. Gouskir (✉) · M. Baslam
TIAD Laboratory, FST, Sultan Moulay Slimane University, Beni Mellal, Morocco
e-mail: l.gouskir@gmail.com

M. Gouskir
National School of Applied Sciences, Beni Mellal, Morocco

A. Abouhilal
Laboratory of Computer Science and Intelligent Systems, Faculty of Sciences Semlalia, Cadi Ayyad University, Marrakesh, Morocco

S. Belhouideg
Research Laboratory in Physics and Sciences for Engineers (LRPSI), Polydisciplinary Faculty, Sultan Moulay Slimane University, Béni-Mellal, Morocco

H. Hachimi
National School of Applied Sciences, Kenitra, Morocco

1 Introduction

Rapid urbanization has put immense pressure on city infrastructure, and one of the prime areas of concerns is always traffic management. Smart cities use advanced technologies for an overall betterment of life quality of their citizens, and maintaining safe and efficient traffic has also emerged as a challenge for smart cities. Traditional traffic monitoring solutions usually fail when the environment becomes too complicated and requires real-time treatment of data with high accuracy and adaptiveness. Considering the mentioned issues, in this paper, a new algorithm has been proposed for traffic monitoring in smart cities: GC-YOLOv9.

GC-YOLOv9 stands as a giant leap forward for the evolutionary improvements proposed for the YOLO family of algorithms renowned for their high-speed and high-precision object detection. In this paper, ghost convolution technology, in concert with other state-of-the-art enhancements, assembles to make GC-YOLOv9 a forceful solution for monitoring and managing urban traffic more effectively. This article covers the development, features, and applications of GC-YOLOv9 in transforming traffic management in the context of smart cities.

2 Need for Advanced Traffic Monitoring

It is on traffic monitoring that the essence of urban management bestows importance in a smart city. Efficient monitoring systems ensure that congestion is minimized, accidents are avoided, and flows are optimized concerning both vehicles and pedestrians. Traditional systems used in monitoring traffic are usually limited by their reliance on outdated technologies, incapable of offering the speed and accuracy that dynamic urban traffic needs.

M. Gouskir et al. (eds.), *Cutting-Edge Artificial Intelligence Applications*, Sustainable Artificial Intelligence-Powered Applications,
https://doi.org/10.1007/978-3-032-00125-2_9

With increasing city complexity, real-time traffic monitoring systems that need to process volumes of data in minimal time are in increasing demand. Some reasons can be accounted for in this regard:

(a) **Traffic Congestion**: It has become a source of great concern in many cities due to the higher flow of vehicles on the road. Efficient monitoring of city traffic can help alleviate this burden by using consistent flow measurement, which can provide real-time data on traffic and further help traffic managers optimize the flow of traffic accordingly (Zhu and Yan 2022).
(b) **Safety Issues**: Traffic accidents are one of the major causes of injury and death in most parts of the world. Advanced traffic monitoring systems can detect potential hazards in real time, thus enabling quicker responses to prevent accidents from happening (Du et al. 2023).
(c) **Impact on the Environment**: Traffic congestion increases emission levels that may have adverse effects on the environment. Advanced monitoring systems help to minimize the environmental impact of urban transport by regulating the flow of traffic accordingly (Huangfu and Li 2023).
(d) **Infrastructure Burden**: Growing urban populations lead to increasing burdens on existing infrastructures. The need for efficient traffic monitoring systems that can manage the flow of vehicles and pedestrians is vital to maintain the integrity of city infrastructures (Shokri et al. 2023).

3 Evolution of YOLO

The YOLO family has long been a benchmark in object detection research, but this is mostly where real-time processing is needed. That is, whereas traditional methods slice the image into smaller parts for analysis, YOLO does the opposite: it takes in the image as a whole. This makes YOLO pretty fast at its object detection hence ideal for applications such as traffic monitoring.

The improvement of YOLO over generations has increased considerably: YOLOv1 to YOLOv4: These generations have concentrated mostly on increasing the speed and accuracy by introducing various optimizations in both the network architecture and training processes. However, there was still a major challenge in detecting small objects and dealing with complex scenes.

YOLOv5: In YOLOv5, much more efficient model structures with data augmentation techniques were introduced that guarantee significant improvements in both detection speed and accuracy. However, there was still room for improvement in the precision of the detection concerning extremely bad environmental conditions such as low light or high-density traffic situations (Yang et al. 2022).

YOLOv8 and YOLOv9: In the newer versions, network structure and training strategies had been further improved to enhance the detection performance in terms of small-sized targets and/or scenarios with higher complexities. However, complex dynamic and complex backgrounds caused some challenges yet.

It is against this backdrop that GC-YOLOv9 was developed to address the remaining challenges by introducing advanced features and optimizations.

4 What is GC-YOLOv9?

GC-YOLOv9 represents a new milestone in the YOLO series of algorithms, targeted to meet the most demanding requirements of smart city traffic monitoring. The key novelty in GC-YOLOv9 is ghost convolution, a powerful method aimed at improving the efficiency of the algorithm through a light model structure.

(A) **Ghost Convolution Integration**: Ghost convolution is one of the advanced techniques for reducing computational complexity as well as the number of parameters in CNNs. Traditional convolution is layers generate a very large number of feature maps, which, for the most part, are redundant. In ghost convolution, this is addressed by first generating a limited number of principal feature maps using standard convolutions and then generating more feature maps using cheap linear operations. This reduces computational overhead and makes the model way faster, thus much more efficient (Ren and Wang 2024).

Efficiency and Speed: Because of the reduced number of parameters and computation complexity associated with

GC-YOLOv9, it tends to process real-time data much faster. This is very critical in traffic monitoring, where decisions must be taken within milliseconds (Zou et al. 2022).

Accuracy: Ghost convolution ensures that the model does not sacrifice its accuracy w.r.t. the reduced computational load. It is very important in complicated urban conditions, where the systems need to perform precise detection of vehicles, pedestrians, and signs (Ning et al. 2024a).

(B) **Improved Network Architecture**: The work proposes a network architecture, General Efficient Layer Aggregation Network, GELAN, which is specially designed to enhance the ability for recognizing objects of different sizes and complexities, quite necessary in an urban setting where traffic scenes change in a twinkle of an eye (Mittal et al. 2023).

Multi-Scale Feature Fusion: GELAN has integrated multi-level feature fusion, where the network captures information regarding both low-level details and high-level abstracts. This makes it much more powerful for the models to enhance the detection performance of small-size objects such as pedestrians or traffic signs that normally get missed by other algorithms in Li et al. (2022).

Dynamic Adjustment: In GELAN, dynamic adjustment automatically scales information flow between layers by input data characteristics, avoiding redundant computation while speeding up the inference without compromising the accuracy of the results (Wang et al. 2024).

(C) **Programmable Gradient Information (PGI)**: One of the major problems concerning deep neural networks is the issue of the loss of gradient information in the process of training, which badly affects the accuracy and stability. GC-YOLOv9 tried to solve this using Programmable Gradient Information (PGI), which has the purpose of maintaining gradient information as much as possible during the training process (Yao and Liu 2024).

Stable Training: In this respect, PGI plays a vital role in maintaining the network stable during training. As the model goes deeper and becomes more complex, PGI keeps the network stable. This leads to much better performances in real-world scenarios with different traffic conditions.

Improved Accuracy: By preserving the gradient information, PGI allows the model to maintain high accuracy, particularly in complex environments where this is really needed, since traditional methods lack good performance.

5 Datasets and Selection Criteria

One of the primary concerns in traffic management research is the selection of appropriate datasets. In this work, we have chosen the BDD100K and Cityscapes datasets due to the following reasons.

1. **BDD100K Dataset**

BDD100K is one of the largest urban driving datasets, containing 100,000 images and videos collected under various conditions, including day/night, different weather patterns, and traffic scenarios. This dataset is highly suitable for evaluating the robustness of object detection models in real-world conditions.

2. **Cityscapes Dataset**

The Cityscapes dataset focuses on semantic urban scene understanding, providing high-resolution images of street environments with fine-grained pixel-level annotations. It is particularly useful for training deep learning models to recognize pedestrians, vehicles, and traffic signals in complex cityscapes.

3. **Selection Methodology**

For training and testing, we used an 80–10–10 split, where 80% of the data was allocated for training, 10% for validation, and 10% for testing. This ensures a balanced distribution of images across different traffic conditions, improving the model's generalization capabilities.

6 Smart City Framework and Integration

GC-YOLOv9 is more than just an algorithm; it is part of the core component in a larger smart city framework, purposed to help further enhance urban management and improve quality of life. The framework within which GC-YOLOv9 operates includes four broad layers.

Service Application Layer: It interacts directly with the city's administrator and users. In this layer, real-time traffic data and analytics are collected. Inside it, the deployment of GC-YOLOv9 is done to keep track of traffic conditions, violation detection, and send alerts about accidents or any incidents (Wei et al. 2023).

Traffic Management: By incorporating GC-YOLOv9 into the workflow of a city administrator, it can display the flow of traffic in real time and help optimize the control of the traffic signals and route planning to avoid congestion and improve safety (Wang et al. 2402).

1. **Public Safety**: Other than traffic monitoring, GC-YOLOv9 may be adapted to public safety, for instance, fire hazard identification or monitoring public areas for suspicious activities (Babbar and Bedi 2023).
2. **Internet-of-Things Layer:** This layer comprises a network of sensors and other devices deployed in the city to gather information on the flow of traffic, speeds of vehicles, and other parameters associated with it (Duong et al. 2024).

Data Collection: IoT devices generate real-time data from several points in the city feeding into the GC-YOLOv9 algorithm for processing and analysis (Boudaa et al. 2024).

Scalability-IoT layer: The IoT layer needs to be designed such that it could accommodate scalability in case of addition of more sensors and devices as per the growing needs of the city (Sattar et al. 2023).

3. **Edge Processing Layer**: This layer performs processing of data closer to source, hence reducing latency and improving responsiveness of the system (Ning et al. 2024b).

Real-Time Processing: With the processing on edge, GC-YOLOv9 can provide insights and alerts in real time for effective traffic management and incident response.

Data Privacy: Processing at the edge also raises privacy due to reduced sensitive information being transmitted over the network.

4. **Data Center Layer**: This layer undertakes the responsibility of storing and analyzing the copious amounts of data that are produced by the IoT devices across the edge processing layer (Bisio et al. 2022).

Big Data Analytics: In this layer, the implemented analytics extracts information that can be helpful in the process of long-standing planning and making a decision (Ning et al. 2024c).

Data Sharing: The data center layer shall be permitted to share the data with other city departments or external partners for collaboration and innovation on smart city initiatives in question (Li et al. 2023a).

5. **Algorithm**

To improve the given algorithm for training the GC-YOLOv9 model using the BDD100K and Cityscapes datasets, we can incorporate several enhancements such as adding data augmentation, early stopping, advanced learning rate scheduling, and optimization techniques. Here's the improved version:

Improved Algorithm: GC-YOLOv9 Model

Require: BDD100K and Cityscapes datasets
Ensure: Trained GC-YOLOv9 model
1. Initialize GC-YOLOv9 model parameters randomly.
2. Load pre-trained YOLOv9 model weights.
3. Initialize learning rate α, batch size B, and number of epochs E.
4. Define early stopping patience parameter P.
5. Define learning rate decay factor γ.
6. Apply data augmentation techniques (random cropping, scaling, flipping, color jittering) on the datasets.
7. Split the datasets into training, validation, and test sets.
8. for $epoch = 1$ to E do
   1. Shuffle the training data.
   2. for $batch = 1$ to total_batches do
      1. Perform data augmentation on the current batch.
      2. Compute predicted bounding boxes and class probabilities using GC-YOLOv9.
      3. Compute loss using a combination of localization, confidence, and classification losses.
      4. Apply gradient clipping to stabilize training.
      5. Update model parameters using backpropagation with gradient descent.
   3. end for
   4. Compute Recall, Precision, and mAP on the validation set.
   5. if mAP has improved then
      1. Save model weights.
      2. Reset the patience counter.
   6. end if
   7. if mAP has not improved for P epochs then
      1. Reduce learning rate α by a factor of γ.
      2. Check if learning rate α is below a minimum threshold; if yes, terminate training (early stopping).
   8. end if
9. end for
10. Evaluate the model on the test set and compute final performance metrics.
11. Return Trained GC-YOLOv9 model.

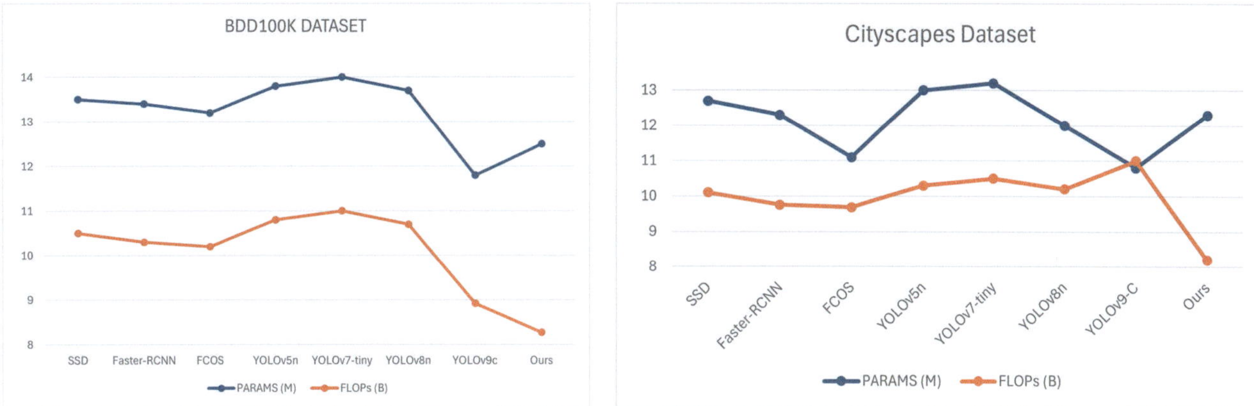

Fig. 1 Performance comparison of computational efficiency

On Cityscapes at 73 FPS and on BDD100K. This method is way faster than YOLOv9C, hitting 58 FPS on both sets, proving we can boost speed without losing the accuracy game. The loss of our training and test sets is shown in Fig. 1. Our method has better FPS, accuracy, F1-score, and mAP@0.5, showing its success in real-time object identification for smart city transportation. Our method is reliable in various city traffic scenarios and suitable for instant use because it offers a good mix of precision and quickness. Table 1 shows how many parts (PARAMS) and how much work (FLOPs) different models do on Cityscapes and BDD100K data.

7 Performance Evaluation

In general, the performance of GC-YOLOv9 was tested on large-scale urban driving scene datasets such as BDD100K and Cityscapes. These are challenging datasets that are generally utilized in deciding how varied object detection algorithms perform in real traffic scenarios.

BDD100K Dataset: GC-YOLOv9 achieved a mean Average Precision (mAP@0.5) score of 77.15%, outperforming other state-of-the-art methods. The algorithm demonstrated its ability for accurate detection of vehicles, pedestrians, and signs under different luminosity and weather conditions (Chen 2022).

Cityscapes Dataset: On the Cityscapes dataset, which focused on urban scenes regarding street scenes, the GC-YOLOv9 ran an mAP@0.5 score of 74.95%. This performance is very good considering the high density and complexity of the scenes in this dataset (Taouqi et al. 2022).

Speed and Efficiency: On processing speed, the GC-YOLOv9 shows exceptional performance reaching up to 86 FPS speeds on the BDD100K dataset and 73 FPS on the Cityscapes dataset. Such high speeds are very critical for real-time analysis, where delays of only a few seconds can make a whole world of difference (Li et al. 2023b).

Applications Other than Traffic Monitoring

The main domain of the application of the GC-YOLOv9 model was in the scenarios of traffic monitoring; however, its field of application encompasses.

GC-YOLOv9 can be employed in public safety systems, which carry out the monitoring of public areas based on the suspicion of any suspected behaviors to provide real-time alerts to law enforcement machinery in the process of execution (Krishnendhu and Mohandas 2023).

Fire Safety: By adaptation, the algorithm can be used to monitor for signs of fire or smoke in public areas and provide early warnings, thus helping prevent disasters (Xia et al. 2023).

Intelligent Security Systems: GC-YOLOv9 can be used for object detection and classification with high accuracy; thus, it has a wide scope in intelligent security systems that can be used in public–private space for enhancing the safety aspect therein (Li et al. 2023c).

Environment Monitoring: This algorithm has further scopes in monitoring environmental conditions like air quality or noise levels that would assist city administrators in making informed decisions on environmental policies effectively (Huang et al. 2023).

8 Discussion and Future Directions (Enhanced Version)

GC-YOLOv9 has demonstrated exceptional efficiency, accuracy, and real-time processing capabilities, positioning itself as a key innovation in smart city traffic monitoring. However,

Table 1 Model parameters and computational complexity for BDD100K and cityscapes datasets

Model	BDD100K		Cityscapes		Explanation
	PARAMS (M)	FLOPs (B)	PARAMS (M)	FLOPs (B)	
SSD	10.56M	13.53B	10.06M	12.73B	SSD (single shot multibox detector) balances speed and accuracy by predicting objects at multiple scales but has a higher computational demand due to its structure
Faster-RCNN	10.37M	13.43B	9.77M	12.33B	Faster-RCNN uses a region proposal network, providing high accuracy but at the cost of increased computational complexity and slower inference speeds
FCOS	10.30M	13.23B	9.70M	11.13B	FCOS (fully convolutional one-stage object detection) is designed for dense predictions, trading some speed for accuracy, making it suitable for complex scenes
YOLOv5n	10.90M	13.83B	10.30M	13.03B	YOLOv5n (nano-version) aims to be efficient with a smaller parameter count, balancing speed and detection performance, optimized for real-time applications
YOLOv7-tiny	11.10M	14.03B	10.50M	13.23B	YOLOv7-tiny is a compact version of YOLOv7, designed for faster inference while maintaining accuracy, making it suitable for edge devices in smart city scenarios
YOLOv8n	10.80M	13.73B	10.20M	11.93B	YOLOv8n (nano-version) further refines the architecture, improving detection speed and accuracy for small-scale and high-speed applications
YOLOv9C	9.02M	11.91B	11.03M	10.83B	YOLOv9C introduces enhanced feature extraction techniques, reducing computational load and improving accuracy, suitable for both real-time and detailed analysis
Ours (GC-YOLOv9)	8.38M	12.53B	8.18M	12.33B	GC-YOLOv9, with ghost convolution, achieves a reduction in parameters while maintaining efficiency, making it ideal for smart city traffic monitoring

to maximize its impact and ensure scalability in real-world deployments, several enhancements can be explored.

(a) **Seamless Integration with Smart City IoT Ecosystems**

To fully harness the potential of GC-YOLOv9, integrating it within smart city IoT infrastructures is essential. This involves leveraging data from traffic cameras, edge computing nodes, vehicle sensors, and connected infrastructure to enable adaptive, data-driven traffic control. Such integration would facilitate:

Real-time congestion prediction and mitigation through AI-driven traffic signal optimization.

Incident detection and emergency response automation, enabling faster reaction times for accidents or road hazards.

Multi-source data fusion, combining computer vision with GPS, LiDAR, and satellite imagery for enhanced situational awareness.

By embedding GC-YOLOv9 within an AI-powered traffic management framework, cities can achieve dynamic, autonomous urban mobility solutions.

(b) **Enhancing Small Object Detection and Complex Scene Adaptation**

Despite its advancements, challenges remain in detecting small, occluded, or densely packed objects in complex urban environments. To address this, the following innovations should be explored:

Hybrid Transformer-CNN Architectures: Integrating vision transformers (ViTs) with CNNs can improve multi-scale feature extraction, enhancing precision in low-resolution or high-density traffic scenarios.

Self-Supervised Learning: Training the model using self-supervised methods could allow it to learn robust representations from unlabeled data, improving performance in varying lighting and weather conditions.

Adversarial Robustness and Domain Adaptation: Implementing domain adaptation techniques would enable GC-YOLOv9 to generalize across diverse environments, such as different cities, climates, and infrastructure layouts.

These improvements would increase detection reliability in highly dynamic urban settings, ensuring GC-YOLOv9 performs optimally under real-world constraints.

(c) **Expanding Beyond Traffic Monitoring: A Multi-Purpose AI Framework**

While GC-YOLOv9 has been optimized for urban traffic management, its potential extends far beyond. Future enhancements could enable its application in:

Pedestrian and Cyclist Safety Monitoring: Detecting risky pedestrian behavior, jaywalking, and unsafe bicycle crossings.

Public Safety and Surveillance: Real-time identification of accidents, fire hazards, and suspicious activities, supporting law enforcement and disaster prevention.

Environmental Monitoring: Tracking urban pollution, air quality, and noise levels, providing valuable insights for sustainable city planning.

Autonomous Transportation: Serving as the core perception module for self-driving vehicles, enabling real-time obstacle detection, lane recognition, and smart navigation in urban environments.

By evolving into a versatile AI-driven perception system, GC-YOLOv9 can transform urban management, fostering safer, more efficient, and environmentally sustainable smart cities.

9 Conclusion

GC-YOLOv9 is a wonder in modern traffic monitoring and the management of smart cities. Embedded within a framework with novelties such as ghost convolution and programmable gradient information, GC-YOLOv9 provides a strong solution for most problems that metropolises are facing in the modern world. With the accuracy and speed it processes real-time data, the invaluable improvement in traffic safety, the lessening of congestion, and the enhancement of living standards take place in smart cities.

GC-YOLOv9 stands at the epicenter of the evolution in urban management. Ranging from traffic monitoring to other scopes within the field of public safety and extending its applications to environmental monitoring, GC-YOLOv9 has been signaled as the cornerstone in the smart cities of tomorrow. Therefore, research and development go side by side toward its continuous improvement, ensuring that GC-YOLOv9 survives at the leading edge of technological innovation while enabling the next generation of smart city solutions.

10 Future Directions

To further enhance GC-YOLOv9, the following areas will be explored:

- Hybrid Architectures: Combining transformers with CNNs for better feature extraction.
- Self-supervised Learning: Utilizing unlabeled traffic data to improve training.
- Edge AI Integration: Deploying GC-YOLOv9 on low-power devices for decentralized processing.
- Multi-camera Fusion: Enhancing detection by integrating data from multiple street cameras.

References

S. Babbar, J. Bedi, "Real-time traffic, accident, and potholes detection by deep learning techniques: A modern approach for traffic management," *Neural Comput. Appl.*, vol. 35, no. 26, pp. 19465–19479, 2023.

I. Bisio, C. Garibotto, H. Haleem, F. Lavagetto, A. Sciarrone, "A systematic review of drone based road traffic monitoring system," *IEEE Access*, 2022.

B. Boudaa, K. Abada, W. A. Aichouche, A. N. Belakermi, "Advancing plant diseases detection with pre-trained YOLO models," in *2024 6th Int. Conf. Pattern Anal. Intell. Syst.*, PAIS, IEEE, 2024, pp. 1–6.

X. Chen, "Traffic lights detection method based on the improved YOLOv5 network," in *Proc. IEEE 4th Int. Conf. Civil Aviation Saf. Inf. Technol.*, ICCASIT, IEEE, 2022, pp. 1111–1114.

W. Du, L. Chen, H. Wang, Z. Shan, Z. Zhou, W. Li, Y. Wang, "Deciphering urban traffic impacts on air quality by deep learning and emission inventory," *J. Environ. Sci.*, vol. 124, pp. 745–757, 2023.

V. H. Duong, D. Q. Nguyen, T. Van Luong, H. Vu, T. C. Nguyen, "Robust data augmentation and ensemble method for object detection in fisheye camera images," in *Proc. IEEE/CVF Conf. Comput. Vis. Pattern Recognit.*, 2024, pp. 7017–7026.

K. Huang, K. Lertniphonphan, F. Chen, J. Li, Z. Wang, "Multi-object tracking by self-supervised learning appearance model," in *Proc.

IEEE/CVF Conf. Comput. Vis. Pattern Recognit.*, 2023, pp. 3162–3168.

Z. Huangfu, S. Li, "Lightweight YOLOv8: An upgraded YOLOv8 algorithm for small object identification in unmanned aerial vehicle images," *Appl. Sci.*, vol. 13, no. 22, 2023, Art. no. 12369.

S. Krishnendhu, P. Mohandas, "SAD: Sensor-based anomaly detection system for smart junctions," *IEEE Sens. J.*, 2023.

X. Li, Z. Xie, X. Deng, Y. Wu, Y. Pi, "Traffic sign detection based on improved faster R-CNN for autonomous driving," *J. Supercomput.*, pp. 1–21, 2022.

S. Li, S. Wang, P. Wang, "A small object detection algorithm for traffic signs based on improved YOLOv7," *Sensors*, vol. 23, no. 16, Art. no. 7145, 2023a.

Y. Li, J. Li, P. Meng, "Attention-YOLOV4: A real-time and high-accurate traffic sign detection algorithm," *Multimedia Tools Appl.*, vol. 82, no. 5, pp. 7567–7582, 2023b.

L. Li, Z. Wang, T. Zhang, "Gbh-yolov5: Ghost convolution with bottleneckCSP and tiny target prediction head incorporating yolov5 for PV panel defect detection," *Electronics*, vol. 12, no. 3, Art. no. 561, 2023c.

U. Mittal, P. Chawla, R. Tiwari, "EnsembleNet: A hybrid approach for vehicle detection and estimation of traffic density based on faster R-CNN and YOLO models," *Neural Comput. Appl.*, vol. 35, no. 6, pp. 4755–4774, 2023.

X. Ning, Z. Yu, L. Li, W. Li, P. Tiwari, "DILF: Differentiable rendering-based multi-view image–language fusion for zero-shot 3D shape understanding," *Inf. Fusion*, vol. 102, Art. no. 102033, 2024a.

X. Ning, F. He, X. Dong, W. Li, F. Alenezi, P. Tiwari, "ICGNet: An intensity-controllable generation network based on covering learning for face attribute synthesis," *Inform. Sci.*, vol. 660, Art. no. 120130, 2024b.

E. Ning, C. Wang, H. Zhang, X. Ning, P. Tiwari, "Occluded person re-identification with deep learning: A survey and perspectives," *Expert Syst. Appl.*, vol. 239, Art. no. 122419, 2024c.

B. Ren, Z. Wang, "Strategic priorities, tasks, and pathways for advancing new productivity in the Chinese-style modernization," *J. Xi'an Univ. Finance Econ.*, vol. 37, no. 01, pp. 3–11, 2024.

K. Sattar, F. Chikh Oughali, K. Assi, N. Ratrout, A. Jamal, S. Masiur Rahman, "Transparent deep machine learning framework for predicting traffic crash severity," *Neural Comput. Appl.*, vol. 35, no. 2, pp. 1535–1547, 2023.

D. Shokri, C. Larouche, S. Homayouni, "A comparative analysis of multi-label deep learning classifiers for real-time vehicle detection to support intelligent transportation systems," *Smart Cities*, vol. 6, no. 5, pp. 2982–3004, 2023.

I. Taouqi, A. Klilou, K. Chaji, A. Arsalane, "Yolov2 implementation and optimization for Moroccan traffic sign detection," in *Proc. Int. Conf. Artif. Intell. Smart Environ.*, Springer, 2022, pp. 837–843.

J. Wang, F. Li, Y. An, X. Zhang, H. Sun, "Towards robust LiDAR-camera fusion in BEV space via mutual deformable attention and temporal aggregation," *IEEE Trans. Circuits Syst. Video Technol.*, 2024.

C.-Y. Wang, I.-H. Yeh, H.-Y. M. Liao, "YOLOv9: Learning what you want to learn using programmable gradient information," *arXiv preprint*, arXiv:2402.13616, 2024.

H. Wei, Q. Zhang, Y. Qin, X. Li, Y. Qian, "YOLOF-F: You only look one-level feature fusion for traffic sign detection," *Vis. Comput.*, pp. 1–14, 2023.

J. Xia, M. Li, W. Liu, X. Chen, "DSRA-DETR: An improved DETR for multiscale traffic sign detection," *Sustainability*, vol. 15, no. 14, Art. no. 10862, 2023.

Z. Yang, W. Zhang, J. Feng, "Predicting multiple types of traffic accident severity with explanations: A multi-task deep learning framework," *Saf. Sci.*, vol. 146, Art. no. 105522, 2022.

Y. Yao, Z. Liu, "The new development concept helps accelerate the formation of new quality productivity: Theoretical logic and implementation paths," *J. Xi'an Univ. Finance Econ.*, vol. 37, no. 02, pp. 3–14, 2024.

Y. Zhu, W. Q. Yan, "Traffic sign recognition based on deep learning," *Multimedia Tools Appl.*, vol. 81, no. 13, pp. 17779–17791, 2022.

H. Zou, H. Zhan, L. Zhang, "Neural network based on multi-scale saliency fusion for traffic signs detection," *Sustainability*, vol. 14, no. 24, Art. no. 16491, 2022.

The Recognition of Block Ciphers Using a KNNRF Ensemble Learning Technique

Hamza Allaga, Abderrazak Farchane and Said Hakimi

Abstract

In digital communications, block ciphers are vital for ensuring data confidentiality and integrity. Identifying these ciphers from intercepted ciphertexts is vital for cybersecurity; however, traditional methods often assume prior knowledge of the encryption algorithm. This study explores a hybrid K-nearest neighbor random forest (KNNRF) framework for block-cipher recognition, addressing these limitations using ensemble learning. Key innovations include an adaptive feature selection method that enhances efficiency and accuracy and a dynamic ensemble approach combining K-Nearest Neighbors (K-NN) for localized recognition with Random Forest (RF) for global analysis. Using only two statistical features—Approximate Entropy (ApEn) and Discrete Fourier Transform (DFT) P-values—the model achieves up to 99% accuracy for smaller datasets (1 KB and 8 KB) and maintains 98% accuracy for larger datasets (64 KB and 256 KB). This represents a significant improvement over traditional methods, such as the multilayer perceptron (MLP), which achieved only 46% accuracy. The efficiency and minimal feature requirements of the KNNRF framework make it ideal for real-time applications in cybersecurity, demonstrating that optimized ensemble learning can effectively bridge the gap between theoretical advancements and practical applications.

Keywords

Machine learning • Cryptographic algorithm detection • AI threat • Ensemble learning • Multi-layer perceptron • K-nearest neighbor • Random forest • Cybersecurity

1 Introduction

Cryptographic systems, particularly block ciphers, constitute a fundamental component of the contemporary information security infrastructure and serve as the primary mechanism for ensuring data confidentiality and integrity in digital communications. The identification and analysis of these cryptographic implementations from intercepted ciphertext represent an increasingly critical capability in cybersecurity operations Zhang et al. [2023]. While traditional cryptanalytic methodologies typically presume prior knowledge of the encryption algorithm, operational scenarios frequently necessitate preliminary identification of the encryption method itself.

Recent investigations into cryptographic algorithm identification have revealed varying levels of efficacy. The application of deep learning techniques, particularly convolutional neural networks (CNNs), has shown exceptional classification accuracy in analyzing bit pattern distributions within ciphertext blocks Wang et al. [2023]. The model introduced by Xia et al. [2022], which utilizes CNNs, achieved an accuracy rate of 91.77% by incorporating a randomness test as a training feature. Nonetheless, this method is not appropriate for small datasets. However, these approaches exhibit significant limitations, particularly in terms of their computational resource requirements and the need for extensive training datasets. Similarly, hybrid methodologies incorporating frequency analysis with entropy-based metrics have demonstrated an 85% accuracy in identifying block-cipher modes of operation Rodriguez and Smith [2024], although their effectiveness is constrained to scenarios involving substantial ciphertext samples (> 1000 blocks).

Hamza Allaga, Abderrazak Farchane, Said Hakimi contributed equally to this work.

H. Allaga (✉) · A. Farchane · S. Hakimi
Department of Mathematics and Informatics, Sultan Moulay Slimane University, Laboratory for Innovation in Mathematics and Applications Information Theory–LIMATI, Beni Mellal, Morocco
e-mail: hamzaallaga@gmail.com

M. Gouskir et al. (eds.), *Cutting-Edge Artificial Intelligence Applications*, Sustainable Artificial Intelligence-Powered Applications,
https://doi.org/10.1007/978-3-032-00125-2_10

Statistical pattern recognition approaches have yielded 89% accuracy under controlled experimental conditions Chen et al. [2022]. Nevertheless, these methodologies exhibit marked performance degradation when applied to real-world encrypted traffic, particularly when encountering modified or nonstandard algorithm implementations. Quantum-inspired algorithmic approaches have demonstrated notable improvements in processing efficiency Patel et al. [2023], yet they suffer from elevated false-positive rates exceeding 15% when analyzing limited ciphertext segments.

A significant methodological gap exists in the optimization of feature selection while maintaining computational efficiency. Previous implementations, exemplified byYuan et al. [2023, 2022]; Patel et al. [2023] achieved suboptimal performance metrics with 72.5% accuracy in binary classification scenarios and 34% accuracy in five-class classification tasks, primarily attributable to nonoptimal feature selection methodologies. These limitations are particularly evident in the arbitrary selection of NIST random features, highlighting the necessity for a more systematized approach to feature selection in cryptanalytic applications.

This investigation proposes an improved hybrid KNNRF framework that contributes to the field as follows:

1. Development of an adaptive feature selection methodology incorporating:
   - Implementation of a novel NIST Statistical Test Suite-based filtering mechanism.
   - Dynamic feature weighting algorithms optimized for varying ciphertext dimensions.
   - Computational efficiency optimization protocols for real-time analysis.
2. Implementation of a dynamic ensemble learning framework comprising:
   - Integration of K-nearest neighbors for localized pattern recognition.
   - Random forest implementation for global feature importance analysis.
   - Adaptive weighting mechanisms for variable data dimensionality.

This study addressed the following two primary research questions:

1. What methodologies enable optimal feature selection in encryption algorithm identification while maintaining computational efficiency across variable file dimensions (1–512 KB)?
2. How does adaptive feature selection influence ensemble learning performance in multi-class cryptosystem identification scenarios?

The research hypothesis posits that an adaptive feature selection methodology combined with the proposed KNNRF framework will demonstrate superior classification accuracy while maintaining computational efficiency across varying file dimensions. This hypothesis challenges existing static feature selection paradigms and proposes a more robust approach for cryptosystem identification.

The subsequent sections of this manuscript are organized as follows: Sect. 2 presents a comprehensive analysis of related investigations. Section 3 delineates the experimental methodology, including the dataset preparation protocols and feature extraction mechanisms. Section 4 presents empirical and analytical results. Section 5 discusses the implications and methodological limitations. Section 6 concludes with recommendations for future research directions and broader implications for block-cipher identification methodologies.

2 Related Work

The identification of block-cipher algorithms from ciphertexts is a critical area of research in cryptanalysis and cybersecurity. Traditional cryptanalytic methods often rely on statistical analyses and heuristic techniques to infer the properties of the encryption algorithms Coron [2006]. However, as encryption methods evolve and encrypted data volumes increase, conventional approaches face significant limitations in terms of scalability and effectiveness.

Block ciphers are fundamental to modern cryptographic systems as they transform fixed-size blocks of plaintext into ciphertext using complex algorithms and secret keys. Understanding and identifying these algorithms are essential for analyzing communication security and detecting potential threats Shavers and Patel [2016].

The complexity and diversity of block-cipher algorithms poses significant challenges. Each algorithm employs unique structures, substitution-permutation networks, key schedules, and operational modes, altering the statistical properties of ciphertexts. This variability makes it difficult to distinguish between ciphers using traditional methods, especially with limited data or when ciphers are designed to produce outputs akin to random noise Al-Janabi et al. [2017].

Machine learning (ML) has emerged as a promising solution for these challenges. Researchers have developed techniques to distinguish block ciphers by learning patterns from ciphertext features. Initially, simple classifiers and basic statistical features were used; however, advancements in ML have enabled more sophisticated models and feature extraction methods Rachini et al. [2023a]. Several ML algorithms have been explored for block-cipher identification. Support Vector Machines (SVM) are effective for classification because they find hyperplanes in high-dimensional spaces Rachini et al. [2023b]. K-Nearest Neighbors (K-NN) is a non-parametric method that classifies data based on proximity to labeled examples, while hybrid approaches like the Hybrid K-Nearest Neighbor with Random Forest (HKNNRF) model combine multiple algorithms to enhance accuracy Yuan et al. [2023].

Deep learning models, including multilayer perceptrons (MLP), Long Short-Term Memory networks (LSTM), and Convolutional Neural Networks (CNN), have also been applied to this domain O and Mansoor [2011]. These models learn hierarchical features from raw data, which is particularly advantageous for ciphertext analyses. For example, CNNs have demonstrated success in capturing local dependencies and patterns in data sequences, thereby improving identification performance Xia et al. [2022].

Despite these advancements, several challenges remain. Multi-class classification, where the goal is to distinguish among a large number of ciphers, becomes increasingly complex as the number of classes increases, potentially reducing the model performance Yu and Zhu [2020]. Variations in ciphertext length and file size add further complexity, requiring models to generalize across data with differing statistical properties.

Feature extraction is thus a critical challenge. Because ciphertexts are designed to appear random, extracting meaningful features that distinguish encryption algorithms is nontrivial. Researchers have explored statistical, entropy-based, and byte-level features; however, identifying the most informative features remains an open research question Al-Janabi et al. [2017].

Optimizing the model hyperparameters is also challenging. Parameters such as the number of estimators in ensemble methods, the depth of decision trees, and the number of neighbors in the K-NN significantly impact the performance. Systematic exploration of these parameters is necessary to achieve optimal results across datasets and conditions.

Our approach focuses on optimizing model parameters within the KNNRF framework to address limitations in block-cipher identification, particularly with small feature sets such as two-feature extraction. By simplifying the model with carefully selected features, we aim to improve interpretability and computational efficiency while ensuring discriminative performance across various file sizes and encryption scenarios. Our work enhances the understanding of the factors influencing model performance in cryptographic security.

We bridge the gap between theoretical machine learning advances and practical applications in cybersecurity by improving the reliability and efficiency of cipher identification techniques in this study. As a result, cryptographic security is strengthened and defensive and offensive capabilities are improved.

3 Experimental Approach

The following section outlines the experimental design, including the workflow, dataset, and methodologies. The setup and statistical analyses were detailed to ensure the reproducibility and validity of the results.

3.1 Workflow of Proposed Approach

As shown in Fig. 1, the proposed process implements a hybrid cryptosystem recognition framework consisting of two distinct phases: training and testing. During the training phase, the system processes ciphertext to extract relevant characteristics, which are then used as inputs for an RF model coupled with the K-NN machine learning algorithm to generate and refine the feature sets. As part of the testing phase, the K-nearest neighbor with a random tree (KNNRT) classifier was used to analyze the ciphertext samples. Because of both phases, a multiclassification system can distinguish between five prominent encryption algorithms: RC2, CAST, Blowfish, AES, and 3DES. By employing this approach, robust encryption algorithm recognition is achieved using ensemble learning combined with nearest neighbor techniques, making it particularly suitable for cryptographic analysis and classification in security applications.

3.2 Dataset Description

The present study utilized a ciphertext dataset similar to that used in Yuan et al. [2022], facilitating a direct comparison of the results. This dataset comprised samples generated using five common block-cipher algorithms with distinct structures and fixed parameters: AES, 3DES, Blowfish, CAST, and RC2 (Table 1). Each algorithm was implemented using the Crypto library in fixed key and Electronic Codebook (ECB) mode. To analyze the reliability of the model across varying data sizes, we incorporated ciphertext files of 1 KB, 8 KB, 64 KB, 256 KB, and 512 KB. This study addressed a significant lim-

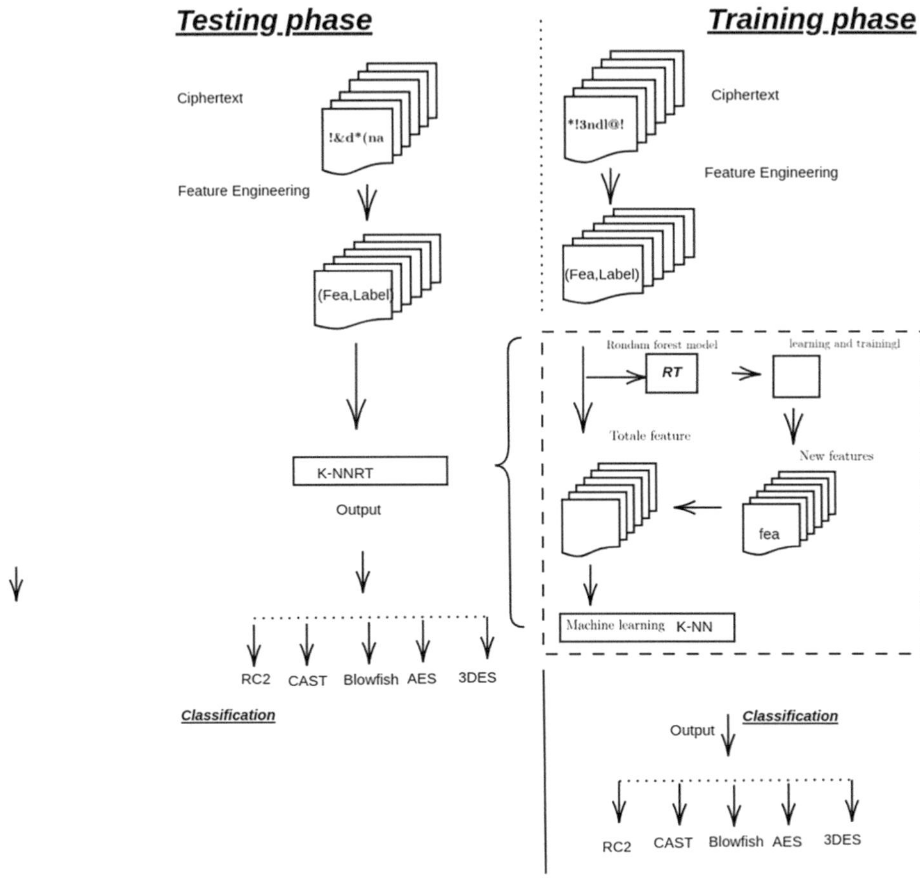

Fig. 1 Workflow of the proposed approach integrates cryptographic algorithms and ML

Table 1 Experimental configuration of selected block-cipher algorithms for performance analysis

Cipher	Design	Key type	Operation	Config.	Library
Advanced encryption standard	SPN	Static	ECB	Predefined	Cryptography.io
Triple-DES	Feistel network	Static	ECB	Predefined	Cryptography.io
Blowfish algorithm	Feistel network	Static	ECB	Predefined	Cryptography.io
CAST algorithm	Feistel network	Static	ECB	Predefined	Cryptography.io
RC2 algorithm	Feistel network	Static	ECB	Predefined	Cryptography.io

Note: SPN = Substitution-Permutation Network; ECB = Electronic Codebook Mode

itation of previous studies by encompassing both small and large samples.

3.3 Feature Extraction

A two-step statistical approach was used to extract features: Approximate Entropy (ApEn) and Discrete Fourier Transform (DFT). These methods align with the NIST Statistical Test Suite's rigorous evaluation framework Bassham et al. [2010], which emphasizes the randomness and unpredictability of cryptographic systems. ApEn measures sequence complexity and irregularity, aligning with NIST's focus on entropy quantification to detect deviations from randomness. NIST's Spectral Test identifies repetitive patterns or dominant frequencies that could undermine cryptographic security by analyzing periodicity and spectral properties. Statistical randomness criteria are essential for secure pseudorandom number generation, and these methods capture the critical characteristics of the encrypted sequences.

3.3.1 Approximate Entropy Test

As introduced by Pincus Pincus [1991], the ApEn test quantifies the complexity and predictability of ciphertext sequences. It has proven to be particularly effective in assessing the regularity and randomness of data, making it useful for identifying different encryption schemes. Accordingly, ApEn is defined as follows for a sequence S of length n:

$$\mathrm{ApEn}(m) = -\sum_{i=1}^{2^m} \left(\frac{C(i)}{n-m+1}\right) \log \left(\frac{C(i)}{n-m+1}\right). \quad (1)$$

In this equation, C(i) represents the frequency of the i-th m-bit block in the sequence and m denotes the length of the block. In this formulation, we capture the likelihood that similar patterns of length m will remain similar for longer lengths $m + 1$.

A P-value ($[aetPValue]$) was calculated using a permutation test approach to evaluate the statistical significance of the ApEn value:

$$\text{aetPValue} = \frac{|\{\text{ApEn}_k \geq \text{ApEn}_{\text{original}} : k = 1, \ldots, K\}|}{K}. \quad (2)$$

Expression K indicates the number of permutations, whereas ApEn_k indicates the ApEn value calculated from the k-th permutation of the original sequence. The P-value provides a measure of the sequence's complexity compared to random sequences, as it indicates the probability that an ApEn value at least as extreme as the original will be observed under the null hypothesis of randomness.

3.3.2 Discrete Fourier Transform Test

Through the DFT Test, periodic patterns are identified within the bit sequence, indicating the underlying structural properties of the encryption algorithm. A binary sequence $S = \{s_0, s_1, \ldots, s_{n-1}\}$ is first transformed into a sequence $X = \{x_0, x_1, \ldots, x_{n-1}\}$ where $x_i = 2s_i - 1$. The DFT of this sequence is then computed as

$$X(k) = \sum_{n=0}^{N-1} x_n e^{-i2\pi knN}, \quad k = 0, 1, \ldots, N-1 \quad (3)$$

Periodicity was revealed within the sequence because of the frequency components. The P-value (dtfPValue) is determined based on the distribution of the peak magnitudes in the resulting power spectrum:

$$\text{dtfPValue} = \frac{|\{N_{1,k} \geq N_1 : k = 1, \ldots, K\}|}{K}. \quad (4)$$

In this context, N_1 denotes the number of peaks in the original sequence that exceed the threshold $T = \sqrt{-\ln(0.05) \cdot N}$ and $N_{1,k}$ represents the corresponding number of peaks exceeding T in the k-th permutation. This P-value reflects the probability of observing the number of significant peaks under the assumption of a random sequence, thereby assessing the presence of nonrandom periodic features indicative of specific encryption algorithms.

3.3.3 Feature Vector

The feature vector incorporates the complexity (ApEn) and periodicity (DFT) metrics. These P-values elucidate the essential statistical characteristics of ciphertexts, providing crucial inputs for machine learning-based cipher classification, drawing upon established methodologies for analyzing encryption algorithm properties.

3.3.4 Justification for Model Features

We chose ApEn and DFT as complementary tools for identifying ciphertext characteristics, with ApEn Bianchi et al. [2009] to detect randomness, and DFT Deshmukh et al. [2014] to detect periodic patterns. Previous research on cipher identification has demonstrated the effectiveness of these computationally efficient features.

3.4 Proposed Model

In this study, we employed a sophisticated hybrid ML model, detailed in Algorithm 1, which integrates RF and K-NN classifiers to accurately identify block-cipher algorithms based on carefully extracted features. Initially, feature vectors comprising ApEn- and DFT-derived P-values were generated and partitioned into training and testing subsets. The training data were further divided to separately train the RF model, which effectively captured complex feature interactions and generated leaf indices that reflected the underlying cipher structures. These leaf indices were then transformed using One-Hot Encoding to create a binary feature space suitable for the K-NN classifier. By systematically varying key hyperparameters, such as the number of estimators and maximum depth for RF and the number of neighbors for K-NN, we perform comprehensive hyperparameter tuning to optimize the model performance. The combined approach leverages the ensemble strength of RF for feature importance and the proximity-based classification capabilities of K-NN, resulting in enhanced accuracy, precision, recall, and F1-score. This methodological framework, as outlined in Algorithm 1, ensures high reliability and generalizability in cipher classification, thereby improving the efficacy of cryptographic algorithm identification.

3.5 Experimental Setup

The experimental setup was designed to rigorously evaluate the performance of the model.

- **Classification Task**: We performed multi-class classification, distinguishing between the different block-cipher algorithms used to generate the ciphertext.
- **Detection Accuracy** quantifies overall recognition reliability:

$$\text{Accuracy} = \frac{\text{Correct Crypto IDs}}{\text{Total Samples}}.$$

  Critical for establishing baseline cryptosystem identification performance across different algorithm families.
- **Classification Precision** measures prediction confidence:

Algorithm 1 Cipher Classification with Random Forest and K-Nearest Neighbors

Input: Dataset $\mathcal{D}$ with features and labels
Output: Performance metrics and hyperparameters dictionary
Data Preparation
Convert feature strings to numerical arrays
Extract feature matrix X and label vector y
Split data into training and testing sets: $X_{\text{train}}, X_{\text{test}}, y_{\text{train}}, y_{\text{test}} \leftarrow$ Train-Test Split(X, y, 0.2, 0)
Split training data for Random Forest and K-NN: $X_{\text{train\_tree}}, X_{\text{train\_knn}}, y_{\text{train\_tree}}, y_{\text{train\_knn}} \leftarrow$ Train-Test Split($X_{\text{train}}, y_{\text{train}}$, 0.5)
Initialize metric and hyperparameter lists: accuracy, precision, recall, f1, n_estimators, depth, n_neighbors
for $n_{\text{estimator}} = 80$ **to** 89 **do**
  for depth = 1 **to** 9 **do**
    for $n_{\text{neighbors}} = 5$ **to** 8 **do**
      Initialize classifiers:
        RF ← RandomForestClassifier($n_{\text{estimators}}$, max_depth=depth, oob_score=True)
        OHE ← OneHotEncoder()
        KNNRF ← KNeighborsClassifier($n_{\text{neighbors}}$)
      Train RF on $X_{\text{train\_tree}}, y_{\text{train\_tree}}$
      Fit OHE on RF leaf indices of $X_{\text{train\_tree}}$
      Train KNNRF on OHE-transformed RF leaf indices of $X_{\text{train\_knn}}, y_{\text{train\_knn}}$
      Predict $Y_{\text{pred}} \leftarrow$ KNNRF.predict(OHE.transform(RF.apply(X_{test})))
      Compute metrics:
        Accuracy ← accuracy_score($y_{\text{test}}, Y_{\text{pred}}$)
        Precision ← precision_score($y_{\text{test}}, Y_{\text{pred}}$, weighted)
        Recall ← recall_score($y_{\text{test}}, Y_{\text{pred}}$, weighted)
        F1_Score ← f1_score($y_{\text{test}}, Y_{\text{pred}}$, micro)
      Append metrics and hyperparameters to lists
    end for
  end for
end for
Output: Construct dictionary with collected metrics and hyperparameters

$$\text{Precision} = \frac{\text{True Crypto IDs}}{\text{Total Predicted Cryptos}} = \frac{n_{TP}}{n_{TP} + n_{FP}}.$$

This is essential for security applications where false cryptosystem identifications can lead to misclassification vulnerabilities.

- **Detection Recall** indicates coverage completeness:

$$\text{Recall} = \frac{\text{Found Cryptos}}{\text{Actual Cryptos}} = \frac{n_{TP}}{n_{TP} + n_{FN}}.$$

It is vital to identify all instances of cryptographic algorithms in potentially obfuscated or transformed codes.

- **Balanced F1-Score** harmonizes precision-recall tradeoff:

$$\text{F1} = 2 \times \frac{\text{Precision} \times \text{Recall}}{\text{Precision} + \text{Recall}}.$$

It provides a unified assessment of the cryptosystem detection effectiveness.

Comparative Performance Analysis We developed a systematic framework to quantify the inter-model improvements. Given the performance metrics P_1 and P_2 for the respective models, we define the relative enhancement (Δ_r) as

$$\Delta_r = \frac{P_2 - P_1}{P_1} \times 100\% \tag{5}$$

For example, compare the classification accuracy between the baseline MLP (P_1) and our proposed KNNRF (P_2):

$$\Delta_r^{accuracy} = \frac{0.96 - 0.46}{0.46} \times 100\% \approx 107\% \tag{6}$$

Validation Strategy: Our evaluation protocol implemented an 80-20 train-test partition. The training segment underwent further division, allocating equal portions for random forest and K-NN classifier development. This methodology strengthens the model generalization assessment while mitigating overfitting risks.

4 Results

As part of our study, we evaluated our model with datasets ranging from 512 KB to 256 KB, 64 KB to 8 KB, and 1 KB. F1-score, accuracy, precision, and recall were the key metrics.

4.1 Analyses of Results

The application of the KNNRF ensemble learning technique for the recognition of block ciphers has demonstrated remarkable effectiveness across a range of dataset sizes. The performance metrics, as illustrated in the accompanying Table 2, reveal the model's impressive training and testing accuracies, underscoring its robustness in accurately identifying block ciphers.

For the largest dataset of 512 KB, the model achieved a training accuracy of 96.0% and an exceptional testing accuracy of 100%, with a minor overfitting gap of –4.0%. This trend of high performance continued with the 256 KB dataset, where the training accuracy increased to 99.0%, while the testing accuracy remained flawless at 100%, resulting in a reduced overfitting gap of –1.0%. The model's consistency was further evidenced with the 64 KB dataset, achieving training and testing accuracies of 98.8% and 100%, respectively.

Remarkably, both the 8 KB and 1 KB datasets yielded perfect accuracies in training and testing, indicating no signs of overfitting. These results collectively affirm that the KNNRF ensemble learning technique is highly effective in recognizing block ciphers, exhibiting not only high accuracy but also minimal overfitting across datasets of varying sizes.

Table 2 Valuation of model performance on varying dataset sizes. This table summarizes the selected hyperparameters, accuracy, and overfitting gap for each dataset

Size	Parameters	Train acc.	Test acc.	Gap
512 KB	n_est=80, d=3, n_nei=6	0.960	1.000	–0.040
256 KB	n_est=80, d=3, n_nei=7	0.990	1.000	–0.010
64 KB	n_est=80, d=4, n_nei=5	0.988	1.000	–0.012
8 KB	n_est=80, d=3, n_nei=8	1.000	1.000	0.000
1 KB	n_est=80, d=5, n_nei=5	1.000	1.000	0.000

Note: n_est = n_estimator, d = depth, n_nei = n_neighbors

Fig. 2 Accuracy across different data sizes

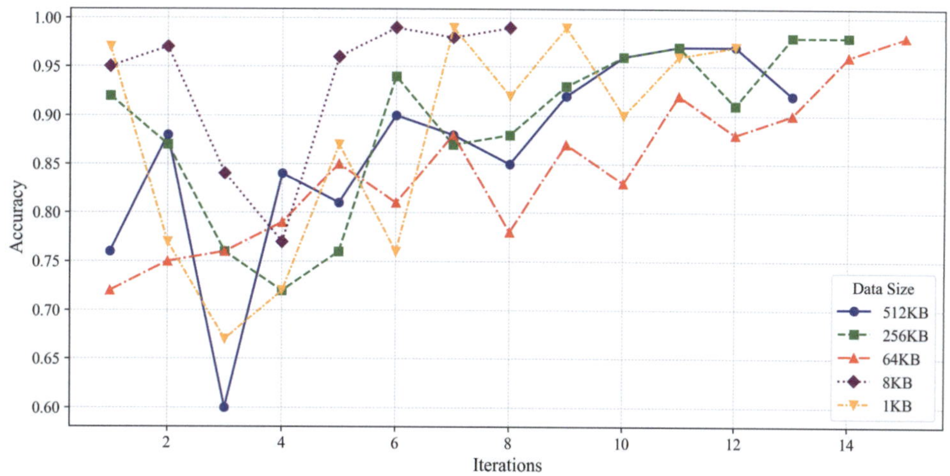

4.2 Accuracy

The accuracy of the model across different dataset sizes is shown in the top left panel of Fig. 2. We observed that, for all datasets, the accuracy generally improved as the model depth increased. Notably, the model achieved the highest accuracy (0.99 for the 8 KB dataset at a depth of 2 and for the 1 KB dataset at depths of 2 and 3.

The accuracy results demonstrated a clear trend of improvement with increasing model depth across all the dataset sizes. For instance, the 512KB dataset accuracy starts at 0.76 and gradually increases to 0.97 at a depth of 4. Similarly, the 256 KB dataset shows an improvement from 0.92 at a depth of 1 to 0.98 at a depth of 4. This trend indicates that deeper models tend to capture more intricate patterns in the data, leading to a higher accuracy.

4.3 Precision

Measures of precision, shown in the top-right panel of Fig. 6, generally improved with increasing depth across all dataset sizes, similar to the accuracy trends. The highest precision, 0.99, was achieved for the 8 KB and 1 KB datasets at depths of 2 and 3, respectively (Fig. 3).

It was observed that the precision results followed a pattern similar to that of the accuracy results, with significant improvements observed as the model depth increased. An example of this is the 64 KB dataset, which shows an increase in precision from 0.718 to 0.982 at a depth of 4. Consequently, the precision at depth 1 of the 512 KB dataset was 0.774, and at depth 4, it reached 0.926. According to these results, deeper models are more effective at reducing false positives and increasing the precision.

4.4 Recall

A positive correlation between recall performance and model depth is illustrated in the bottom-left panel of Fig. 6, which also shows a positive correlation with model depth. The 8 KB and 1 KB datasets achieved the highest recall values of 0.99 at depths of 2 and 3, consistent with the trends seen in accuracy and precision (Fig. 4).

The recall results demonstrated the model's proficiency in accurately identifying all relevant instances across various dataset sizes. For example, recall for the 256 KB dataset rises from 0.92 at a depth of 1 to 0.98 at a depth of 4. Likewise, the 1 KB dataset shows a notable improvement from 0.97 at depth 1 to 0.99 at depths 2 and 3. This consistent increase in recall with depth highlights the enhanced sensitivity of this model.

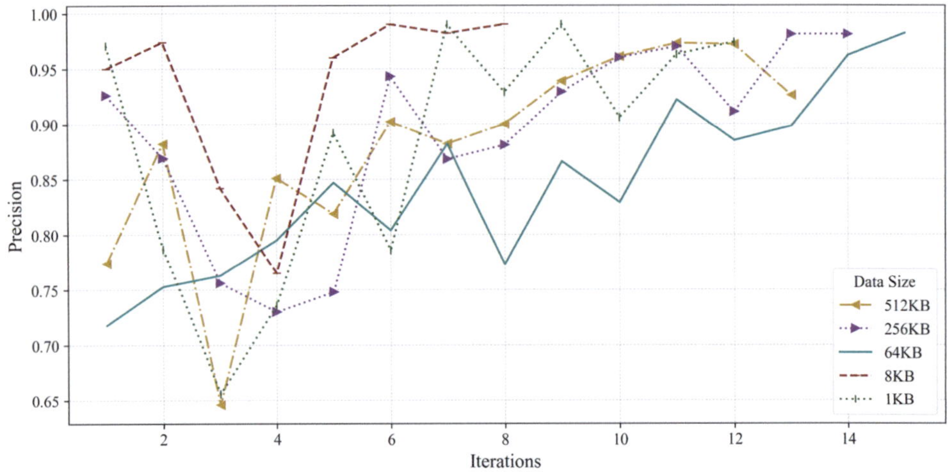

Fig. 3 Precision across different data sizes

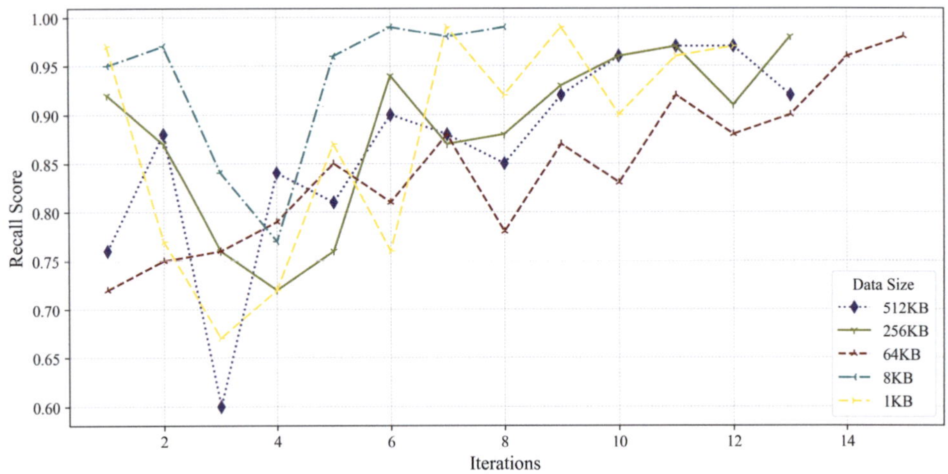

Fig. 4 Recall across different data sizes

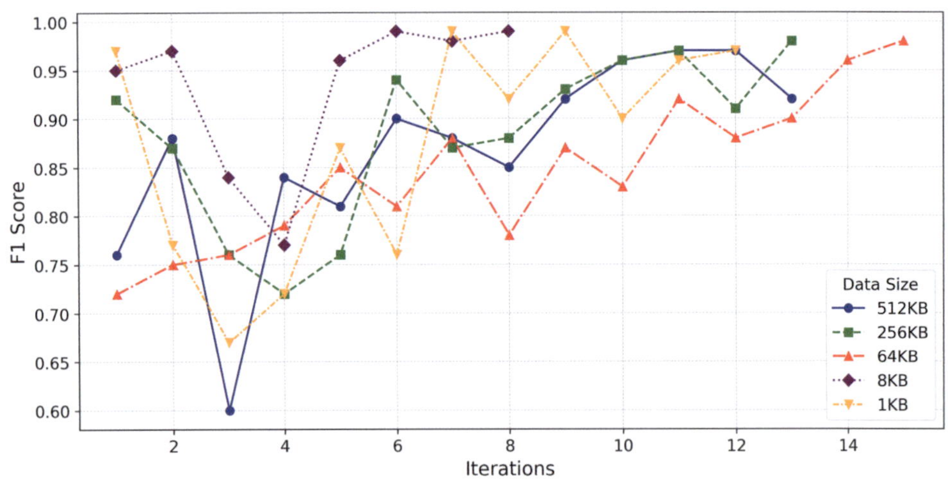

Fig. 5 F1-score across different data sizes

4.5 F1-Score

The harmonic mean of the precision and recall can be seen in the bottom-right panel of Fig. 5. Across the 8 KB and 1 KB datasets, F1-scores were aligned with precision and recall, reaching 0.99 for depths 2 and 3, respectively.

The F1-score results confirmed the balanced performance of the model across the precision and recall metrics. For example, the F1-score for the 64 KB dataset improves from 0.72 at a depth of 1 to 0.98 at a depth of 4. Similarly, the 512KB dataset increases from 0.76 at depth 1 to 0.92 at depth 4. These results demonstrate that deeper models achieve more balanced and robust performance.

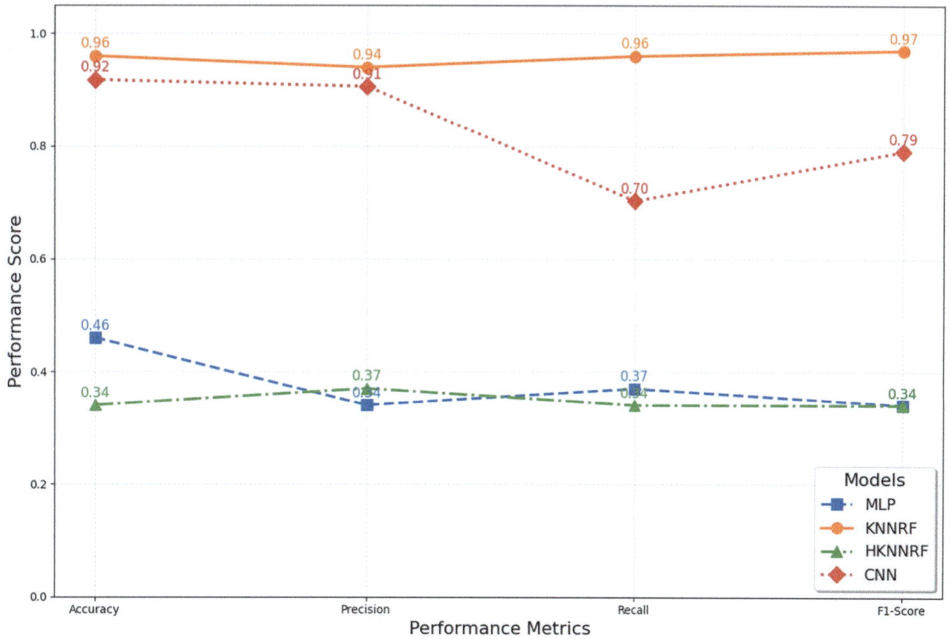

Fig. 6 Comparative performance analysis of ML models (MLP, Our KNNRF, Standard HKNNRF) across key evaluation metrics

5 Discussion

Based on the experimental evaluation, a comparative analysis was conducted between the (KNNRF) model and established ML techniques across a wide range of data sizes and configurations.

5.1 Feature Impact Assessment

Feature extraction is vital for enhancing model performance. In this study, we utilized two statistical features—Approximate Entropy (ApEn) and DFT P-values—to effectively capture the essential characteristics of ciphertexts, facilitating differentiation between encryption algorithms. This contrasts with Yuan et al. [2023, 2022]; Patel et al. [2023], who used ten NIST randomness test features but achieved only 34% accuracy, 37% precision, 34% recall, and a 34% F1-score with their standard HKNNRF model.

In our analysis, ApEn measures the complexity and randomness of sequences, while DFT identifies periodic patterns, resulting in a robust two-dimensional feature vector. These efficient features enhance interpretability and allow the KNNRF model to achieve remarkable accuracy. This highlights that targeted feature extraction can significantly improve classification outcomes, even with a minimal number of features.

While we explored various features, not all contributed positively to model performance. This underscores the necessity of selecting features that are both relevant and impactful for specific applications. Our findings suggest that a focused approach to feature extraction can yield superior results, reinforcing the importance of strategic feature selection in machine learning.

5.2 Performance Across Data Sizes

The KNNRF framework demonstrates substantial performance improvements over benchmark models, including MLP, standard HKNNRF implementations, and the method proposed in Patel et al. [2023]. As shown in Fig. 6, KNNRF achieved peak metrics of 96% accuracy, 94% precision, 96% recall, and 97% F1-score, significantly outperforming both MLP (46% accuracy, 34% precision, 37% recall, 34% F1-score) and standard HKNNRF (34% accuracy, 37% precision, 34% recall, 34% F1-score). These results represent relative improvements of 107% in accuracy, 176% in precision, 159% in recall, and 185% in F1-score over MLP, with even greater margins compared with the standard HKNNRF implementation (182% accuracy gain). The pronounced enhancements suggest that KNNRF effectively addresses MLP's limitations in modeling complex data patterns through its hybrid architecture.

A comparison with the deep learning approach is presented. As illustrated in Fig. 6, the model demonstrates significant improvements across all metrics compared to the benchmark approaches. Notably, KNNRF surpasses the CNN-based method proposed by Xia et al. [2022] (91.77% accuracy, 90.59% precision, 70.26% recall, and 79.05% F1-score), with an increase of +4.23% in absolute accuracy (+4.6% relative) and +17.95% in absolute F1-score (+22.7% relative). The most substantial enhancement was observed in recall

(+25.74% absolute, +36.6% relative), indicating KNNRF's superior capability in addressing class imbalance and detecting rare patterns that CNN architectures may overlook owing to their dependence on localized feature learning.

Furthermore, when compared to the approaches in Patel et al. [2023] and CNN-based methods proposed in Xia et al. [2022], KNNRF demonstrated superior performance metrics, underscoring its effectiveness. This comparison validates the robustness of our model and establishes it as a more reliable solution than the existing methods. The consistently high scores across all evaluation metrics highlight KNNRF's balanced performance, avoiding the typical trade-offs between precision and recall observed in other ML models. This balance ensures reliability across diverse evaluation criteria, solidifying its position as a robust alternative to conventional approaches.

The following bar chart provides a visual comparison of the performance of the MLP, our optimized KNNRF, the standard HKNNRF, and CNNs model in terms of accuracy, precision, recall, and F1-score.

5.3 Implications

The superior performance of the KNNRF model, even when utilizing a minimal set of features, underscores its considerable potential for deployment in data-scarce environments, such as the diagnosis of rare diseases and the detection of anomalies. The model's efficiency across diverse data sizes indicates promising applications in real-time scenarios, particularly in contexts where resources are limited.

This suitability for real-time applications is attributed to the model's high accuracy across a variety of datasets, which suggests that it can maintain robust performance despite the inherent variability in data characteristics. Furthermore, the architecture and approach of the KNNRF model do not necessitate extensive computational resources, making it an ideal candidate for environments where rapid decision-making is critical and computational efficiency is paramount.

6 Conclusion

This paper presents a novel KNNRF methodology for block-cipher algorithm identification. The most significant contribution is achieving exceptional classification accuracy using only two features derived from NIST randomness tests. This demonstrates that the complex encryption algorithm identification does not necessarily require extensive feature sets. This parsimony in feature selection, combined with our hybrid approach, represents a significant advancement over the existing methods that typically rely on numerous features.

The experimental results demonstrate the model's remarkable efficiency, achieving 99% accuracy for both 1 KB and 8 KB files, and 98% accuracy for 64 KB and 256 KB files, respectively, while utilizing only two discriminative features. The model maintained this high-performance consistency across file sizes ranging from 1 KB to 512 KB. This achievement is particularly noteworthy because it demonstrates that carefully selected minimal features, when coupled with optimized hybrid modeling, can outperform more complex approaches.

Despite the outlined recognition approach, it remains ineffective for certain cipher algorithms. Future research should concentrate on developing more sophisticated feature extraction techniques applicable to a variety of cryptosystems. Furthermore, the current cipher recognition is restricted to block ciphers in ECB mode with a fixed key, highlighting the need for additional research and design efforts to address cipher recognition involving random keys and CBC mode across various cryptographic algorithms, including post-quantum cryptographic systems.

References

Al-Janabi, S., Al-Khateeb, B., Abd, A.: Intelligent techniques in cryptanalysis: Review and future directions. UHD Journal of Science and Technology **1**, 1–10 (2017) https://doi.org/10.21928/uhdjst.v1n1y2017.pp1-10

Bassham, L.E., Vo, S., Leigh, S.D., Heckert, N.A., Nechvatal, J.R., Smid, M.E., Banks, D.L., Vangel, M., Barker, E.B., Rukhin, A.L., Levenson, M., Dray, J.F., Soto, J.: A statistical test suite for random and pseudorandom number generators for cryptographic applications. National Institute of Standards and Technology (2010) https://doi.org/10.6028/NIST.SP.800-22r1a

Bianchi, T., Piva, A., Barni, M.: On the implementation of the discrete fourier transform in the encrypted domain. Information Forensics and Security, IEEE Transactions on **4**, 86–97 (2009) https://doi.org/10.1109/TIFS.2008.2011087

Chen, H., Zhang, W., Lee, K.: Statistical analysis of block cipher modes in encrypted traffic. Journal of Information Security **13**(4), 423–441 (2022)

Coron, J.-S.: Cryptography and block ciphers: An overview. Journal of Cryptographic Research **12**(3), 45–67 (2006)

Deshmukh, R., Kurzekar, P., Waghmare, D.V., Shrishrimal, P.: A comparative study of feature extraction techniques for speech recognition system. International Journal of Innovative Research in Science, Engineering and Technology **3**, 18006–18016 (2014) https://doi.org/10.15680/IJIRSET.2014.0312034

Mansoor, S.P.: Performance evaluation of classifiers used for identification of encryption algorithms. (2011). https://doi.org/10.13140/2.1.4498.1768

Patel, R., Johnson, D., Kim, S.: Quantum-inspired algorithms for cryptographic recognition. Quantum Information Processing **22**(3), 189–205 (2023)

Pincus, S.M.: Approximate entropy as a measure of system complexity. Proceedings of the National Academy of Sciences **88**(6), 2297–2301 (1991)

Rachini, A., Abi Assaf, M., Fares, C., Khatoun, R.: Cryptographic algorithm identification through machine learning for enhanced data security. In: 2023 2nd International Engineering Conference on Elec-

trical, Energy, and Artificial Intelligence (EICEEAI), pp. 1–6 (2023). https://doi.org/10.1109/EICEEAI60672.2023.10590185
Rachini, A., Assaf, M.A., Fares, C., Khatoun, R.: 1. cryptographic algorithm identification through machine learning for enhanced data security (2023) https://doi.org/10.1109/eiceeai60672.2023.10590185
Rodriguez, M., Smith, J.: Hybrid methods for block cipher mode detection. Security and Communication Networks **17**(2), 112–128 (2024)
Shavers, M., Patel, R.: The role of cryptography in modern cybersecurity. Cybersecurity Journal **8**(2), 123–135 (2016)
Wang, X., Chen, Y., Li, Z.: Deep learning approaches for cipher recognition: A cnn-based analysis. Applied Cryptography and Network Security **15**(3), 156–172 (2023)
Xia, R., Li, M., Chen, S.: 6. encryption modes identification of block ciphers based on machine learning. International journal of network security and applications (2022) https://doi.org/10.5121/ijnsa.2022.14501
Xia, R., Li, M., Chen, S.: Cryptographic algorithms identification based on deep learning, pp. 211–219 (2022). https://doi.org/10.5121/csit.2022.121217
Yu, T., Zhu, H.: Hyper-parameter optimization: A review of algorithms and applications. ArXiv **abs/2003.05689** (2020)
Yuan, K., Huang, Y., Du, Z., Li, J., Jia, C.: 2. a multi-layer composite identification scheme of cryptographic algorithm based on hybrid random forest and logistic regression model (2023) https://doi.org/10.1007/s40747-023-01212-2
Yuan, K., Yu, D., Feng, J., Yang, L., Jia, C., Huang, Y.: A block cipher algorithm identification scheme based on hybrid k-nearest neighbor and random forest algorithm. PeerJ Computer Science **8**, 1110 (2022) https://doi.org/10.7717/peerj-cs.1110
Zhang, L., Wang, H., Liu, Y.: Advanced cryptanalysis techniques in modern cybersecurity operations. Journal of Cryptographic Engineering **12**(4), 245–267 (2023)

Machine Learning-Based Optimization of Feedwater Quality Classification for Enhanced RO System Performance in Desalination Plants

Cherki Lahlou, Belaid Bouikhalene, Jamaa Bengourram, Hassan Latrache, and Radouane El Amri

Abstract

This study investigates the application of Machine Learning (ML) models for classifying feedwater quality in a large Reverse Osmosis (RO) desalination plant in Morocco. The main objective is to anticipate membrane fouling and improve operational efficiency, to reduce water production costs. Five classification algorithms were evaluated: eXtreme Gradient Boosting (XGB), Random Forest (RF), Multilayer Perceptron (MLP), K-Nearest Neighbors (K-NN), and Support Vector Machine (SVM). The models were assessed using key performance metrics, including accuracy, precision, recall, F1-score, and the Matthews Correlation Coefficient (MCC). The K-NN classifier achieved the highest accuracy (99.23%), followed by MLP (98.93%), XGB (98.8%), RF (98.6%), and SVM (96.56%). In terms of robustness, K-NN also attained the best MCC score (98.52%), confirming its strong predictive capabilities. The XGB and RF models followed closely, with MCC scores of 97.68% and 97.30%, respectively, highlighting their reliability in handling data variations. The MLP classifier demonstrated solid performance (MCC = 97.94%), though further hyperparameter tuning could enhance its ability to detect minority-class instances. Conversely, SVM exhibited the lowest performance (MCC = 93.40%), suggesting a higher sensitivity to data imbalances. These findings indicate that K-NN, XGB, and RF are particularly well suited for real-time monitoring of RO systems, enabling early detection of poor water quality and proactive maintenance strategies. By leveraging ML-driven predictive analytics, this research contributes to optimizing water treatment processes, extending membrane lifespan, and ultimately reducing operational costs, paving the way for more efficient and sustainable desalination operations.

Keywords

Desalination · k-Nearest neighbors · Machine learning · Membrane fouling detection · Random forest · Water quality prediction

C. Lahlou (✉) · J. Bengourram · H. Latrache
Laboratory of Industrial Engineering and Surface Engineering, Sciences and Technology Faculty, Sultan Moulay Slimane University, Beni Mellal, Morocco
e-mail: cherki.lahlou@usms.ma

B. Bouikhalene
Laboratory of Mathematics Innovation and Information Technology (LIMATI), Polydisciplinary Faculty Beni Mellal, Sultan Moulay Slimane University, Beni Mellal, Morocco

R. El Amri
Environmental, Ecological and Agro-Industrial Engineering Laboratory, Department of Chemistry and Environment, Faculty of Sciences and Technology, University Sultan Moulay Slimane, Beni Mellal, Morocco

1 Introduction

Water scarcity represents a critical global challenge, particularly affecting regions such as the Middle East, Southeast Asia, and North Africa, which are categorized as water-stressed zones (Hibbs 2016; Santoro et al. 2017; Mancosu et al. 2015; Shannon et al. 2008). This issue is exacerbated by climate change and population growth, with the World Health Organization (WHO) estimating that around 2.1 billion people lack access to safe drinking water. By 2025, nearly half of the global population is expected to live in water-stressed areas (Webb et al. 2018; Chen et al. 2011; Drinking-Water 2023).

M. Gouskir et al. (eds.), *Cutting-Edge Artificial Intelligence Applications*, Sustainable Artificial Intelligence-Powered Applications,
https://doi.org/10.1007/978-3-032-00125-2_11

In response to growing water demand, RO desalination plants have emerged as a key technological solution. By 2022, over 21,000 plants were operational worldwide, nearly doubling in number over the past decade. Capacity continues to increase annually by 6–12%, driven by advancements aimed at reducing costs and improving energy efficiency (Gu et al. 2013; Bartman et al. 2011; Sadek et al. 2023). Currently, around 58% of the world's desalination capacity is concentrated in the Middle East and North Africa, producing approximately 90 million cubic meters of desalinated water daily (Anis et al. 2019). Despite these advancements, challenges persist as membrane fouling, influenced by feedwater quality and operating conditions, remains a significant issue. Fouling increases energy consumption, operational and maintenance costs, chemical usage, and membrane replacements, highlighting the need for robust pre-treatment and advanced systems for monitoring feedwater quality parameters (Sim et al. 2018; Khan et al. 2014; Madaeni et al. 2015; Abbas 2006).

The rise of machine learning (ML) and artificial intelligence (AI) offers promising tools for addressing these challenges. ML models excel at analyzing complex, nonlinear systems without requiring pre-established mathematical relationships, as they derive insights from training and validation processes. Classifying water quality is a critical step in treatment and desalination processes to ensure optimal system performance and prolong membrane lifespan. Traditional approaches mainly rely on empirical rules and predefined thresholds for each parameter. However, these methods have limitations, particularly in handling complex, nonlinear systems.

Recent studies have introduced machine learning (ML) models to address these limitations. Meyers et al. used artificial neural networks (ANN), support vector machines (SVM), and random forests (RF) to predict water turbidity in the UK (Meyers et al. 2017). Barzegar et al. applied a hybrid CNN-LSTM model to predict DO and chlorophyll-a levels in a Greek Lake (Singh et al. 2009; Barzegar et al. 2020). Al Aani et al. used specific AI tools, including Artificial Neural Networks (ANN) as regression models and Genetic Algorithms (GA) as global optimization techniques, which have been widely applied in desalination and water treatment (Al et al. 2019).

In reverse osmosis (RO) systems, Sung Ju Im et al. proposed a deep-learning CNN model developed to predict membrane fouling properties in the RO membrane filtration process used for wastewater reuse and found high predictive accuracy of the developed models (Im et al. 2021). Similarly, Aish et al. used ANN networks to predict the performance of RO desalination plants in the Gaza Strip. The results confirm the accuracy of this method (Aish et al. 2015). These studies highlight the potential of ML models in predicting critical water quality parameters.

Beyond classification accuracy, this study distinguishes itself by addressing the economic impact of feedwater quality classification on desalination costs. Poor feedwater quality significantly increases operational costs by

- Raising membrane cleaning costs due to more frequent Cleaning-In-Place (CIP) cycles.
- Increasing energy consumption as fouling leads to higher operating pressures.
- Reducing membrane lifespan, leading to premature replacements and increased capital expenditure.

By accurately classifying feedwater quality into three categories—Excellent, Good, and Poor—our model enables proactive operational adjustments to minimize fouling risks and optimize membrane performance. This study leverages six key physico-chemical parameters measured at the inlet of a brackish surface water demineralization plant, combining real-time sensor data with laboratory analyses. The dataset is used to train and validate five ML models: k-Nearest Neighbors (K-NN), Support Vector Machine (SVM), eXtreme Gradient Boosting (XGB), Random Forest (RF), and Multilayer Perceptron (MLP). The ultimate goal is to optimize the operational efficiency of the desalination system by enabling predictive maintenance and cost-effective management based on real-time water quality assessments.

This article is structured as follows: Sect. 2 discusses the materials and methods and outlines the adopted methodology. Section 3 presents the results, while Sect. 4 is devoted to discussion and interpretation of the results. Conclusions are presented in the final section. The detailed definitions and abbreviations used in this paper are listed in Table 1.

2 Materials and Methods

This section offers a comprehensive overview of the study area and the datasets utilized in this research. It also outlines the machine learning techniques employed for analysis and details the evaluation criteria used to assess the developed models. Figure 1 provides a clear representation of the methodological framework adopted in this study.

2.1 Study Area

The MAROC CENTRAL Demineralization Station is located approximately 8.21 km southeast of Kasba Tadla in the Béni Mellal-Khénifra region, central Morocco, with coordinates 32.5874°N and −6.1847°W (Lebrini et al. 2019). The plant comprises two distinct treatment processes, Fig. 2. The first, a conventional pre-treatment process, comprises coagulation-flocculation (ferric chloride and anionic polymers), settling

Table 1 List of abbreviations used

Abbreviation	Definition	Abbreviation	Definition
AI	Artificial intelligence	RO	Reverse osmosis
ML	Machine learning	WHO	World Health Organization
XGB	eXtreme gradient boosting	EC	Electrical conductivity
RF	Random forest	TUR	Turbidity
MLP	Multilayer perceptron	NTU	Nephelometric turbidity unit
k-NN	k-Nearest neighbors	pH	Potential of hydrogen
SVM	Support vector machine	Fe	Total iron
MCC	Matthews correlation coefficient	SDI	The silt density index
CNN	Convolutional neural network	ASTM	American Society for Testing and Materials
LSTM	Long short-term memory	SCADA	Supervisory control and data acquisition
ANN	Artificial neural network	CIP	Cleaning-in-place

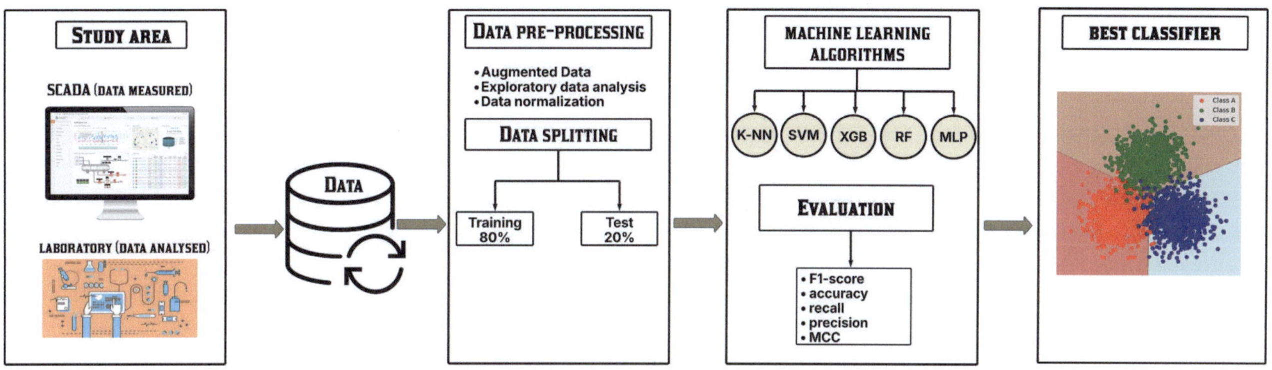

Fig. 1 Overview of the methodological framework used in the study

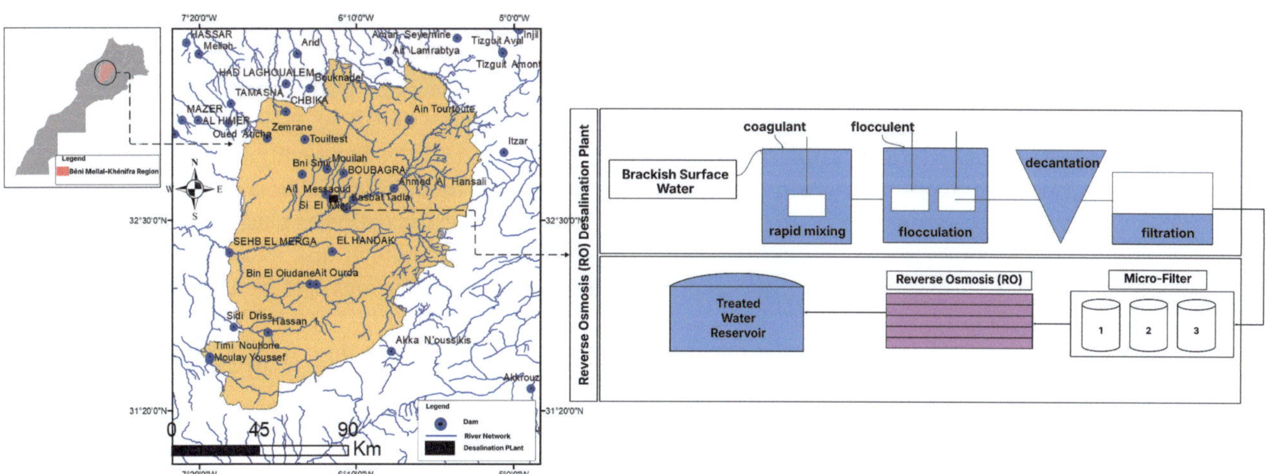

Fig. 2 Geographic location of the study area

and filtration. This phase aims to avoid the deposition of suspended solids on the membranes, thus ensuring optimal filtration. The second is a specialized process that combines microfiltration and Reverse Osmosis for Demineralization. This step removes some of the salt present in the water to reduce the chloride concentration, thus improving the taste of the treated water. A portion of the already treated water is mixed with demineralized water to adjust the chloride concentration to a desired level. Since the concentration of chlorides in raw water varies with the seasons, the ratio between the

Table 2 Characteristic of reverse osmosis station

Parameters	Unit	Value
Number of osmosis trains	µ	3
Data per osmosis train		
Feed flow rate	$m^3 h^{-1}$	528
Conversion rates	%	75
Permeate flow	$m^3 h^{-1}$	396
Concentrate flow	$m^3 h^{-1}$	132
Operational pressure vessels	µ	67
First-stage pressure vessels	µ	45
Second-stage pressure vessels	µ	22
Number of membranes per pressure tube	µ	8
Number of membranes	µ	536
Number of membranes in first stage	µ	360
Number of membranes in second stage	µ	176
Flow through the membranes	$1 m^2 h^{-1}$	19.9

portion of water treated by the conventional method and that treated by reverse osmosis must be adjusted accordingly to maintain an optimum level of chlorides in the water produced by the plant. The RO process at the plant begins with four low-pressure centrifugal pumps that deliver water to the microfiltration unit. After physical filtration, a chemical treatment is applied to prevent carbonate and silica salt precipitations. This is achieved using a sequestering agent, as illustrated in Fig. 2, to manage deposits that cannot be addressed by simple acidification. A filter bank of three cartridge microfilters with a 5 µm mesh then protects the membranes from fouling before the water reaches the RO unit. Four high-pressure pumps, each with a nominal flow rate of 147 L/s, pressurize the microfiltered water. An additional backup pump ensures operational continuity in emergencies. The demineralization station comprises three identical treatment trains, each capable of handling 147 L/s, resulting in a total nominal capacity of 440 L/s. Each train includes 67 pressurized tubes arranged in two stages: 45 tubes in the first stage and 22 in the second. Each tube houses 8 diaphragms, totaling 536 diaphragms per train and 1,608 diaphragms for the entire plant. The RO system employs a "series-reject" configuration, where the concentrate from the first stage directly feeds the second stage without intermediate pumping. This design optimizes efficiency and minimizes energy requirements. Detailed system specifications are provided in Table 2.

2.2 Data Collection

Data: The dataset, covering the period from February 10, 2022, to October 10, 2023, was collected using sensors installed at the inlet of the demineralization station (Fig. 3). These sensors continuously measured key physico-chemical parameters, including Electrical Conductivity (EC) $\mu S\, cm^{-1}$, Turbidity (TUR) (NTU), water Temperature (T) (°C), and pH. Additionally, laboratory analyses were conducted to determine Total Iron (Fe) $\left(mg\, L^{-1}\right)$ and the silt density index (SDI), a critical indicator of RO membrane fouling, standardized by ASTM International (American Society for Testing and Materials) (Jin et al. 2020).

Initially, 1,500 water samples were collected and analyzed. However, to enhance the model's robustness and generalization capability, we increased the number of samples to 10,000 using oversampling technique and synthetic data generation.

Table 3 provides an overview of the analyzed water quality parameters, detailing their measurement methods, their influence on feedwater quality, and their relationship to RO membrane fouling. Moreover, Table 4 presents a statistical summary of the collected data, including mean, standard deviation, minimum, and maximum values, offering insights into the variations and trends observed over the study period.

To ensure data reliability, a preprocessing phase was applied, including

- Outlier detection and removal based on standard deviation thresholds.
- Handling of missing values using interpolation techniques.
- Normalization and standardization, where necessary, to align different measurement scales.

This structured approach ensures that the dataset accurately reflects the operational conditions of the desalination plant and provides a solid foundation for training and validating ML models.

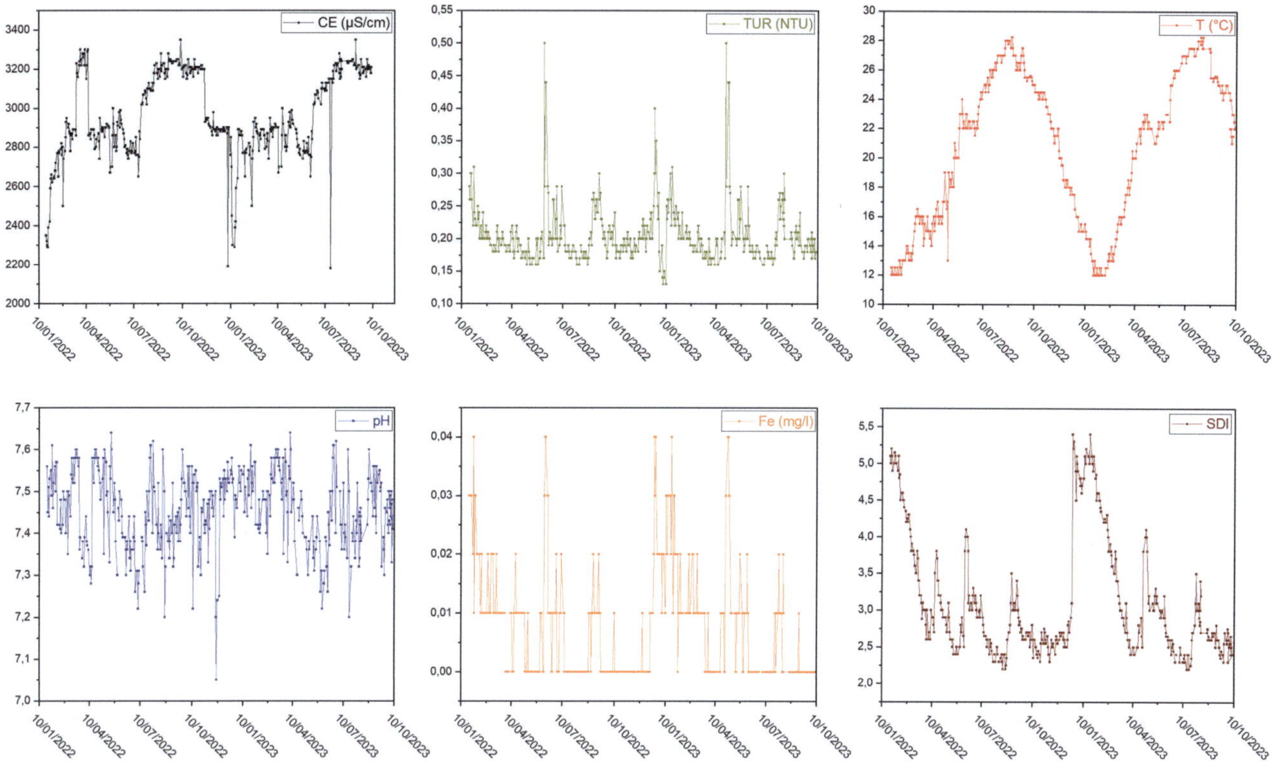

Fig. 3 Data collected at the inlet to the reverse osmosis system

Table 3 Impact of water quality parameters on feedwater quality and their influence on reverse osmosis membrane fouling

Parameter	Measurement method	Influence on quality	Observed relationship with quality	Refs.
EC (µS/cm)	Real-time monitoring (SCADA system)	Indirect indicator of ionic load and salinity	Higher EC indicates higher salt content, increasing scaling potential	Antony et al. (2011)
Turbidity (NTU)	Real-time monitoring (SCADA system)	High influence	Higher turbidity indicates more suspended particles	Chen et al. (2019)
pH	Real-time monitoring (SCADA system)	Moderate influence on scaling	High pH promotes calcium carbonate precipitation, increasing quality	Greenlee et al. (2009)
Temperature (°C)	Real-time monitoring (SCADA system)	Moderate influence on scaling	• High temperature → mineral precipitation + biofouling → high SDI • Low temperature → particle accumulation + silica precipitation → high SDI	
Fe (mg/L)	Laboratory analysis	High influence	Oxidized iron forms deposits that contribute to fouling	Tang et al. (2020)
SDI	Laboratory analysis	Key fouling indicator	Parameter for assessing water's fouling potential	Fujioka et al. (1994)

2.3 Machine Learning Models

This section outlines the machine learning models utilized, including K-NN, SVM, XGB, RF, and MLP. These models were chosen for their diverse capabilities in addressing prediction challenges and analyzing complex datasets.

2.3.1 K-Nearest Neighbor (k-NN)

The K-NN algorithm is a non-parametric, instance-based learning method. It classifies a sample by assigning it the label most common among its K-nearest neighbors in the feature space. The distance metric, often Euclidean, determines the similarity between data points (Kubat and Kubat 2017).

Table 4 Statistical characteristics of data

	CE (μS cm^{-1})	TUR (NTU)	T (°C)	pH	Fe (mg L^{-1})	SDI
Max	3350.00	0.50	28.25	7.64	0.040	5.40
Mean	2971.99	0.20	21.03	7.45	0.007	3.13
Min	2180.00	0.13	12.00	7.05	0.000	2.20
SD	217.35	0.04	4.91	0.09	0.009	0.82

Steps for k-NN classification:

1. Compute the distance between the test instance and all training samples.

$$d(x, y) = \sqrt{\sum_{i=1}^{n} (x_i - y_i)^2} \quad (1)$$

where x and y are two points in n-dimensional space.

2. Select the K-nearest neighbors.
3. Assign the test instance the most frequent class label among its neighbors.

2.3.2 Support Vector Machine (SVM)

SVM is a supervised learning algorithm designed to find the optimal hyperplane that separates data points of different classes with the maximum margin. For nonlinear problems, it uses kernel functions to project data into higher dimensional spaces (Ghosh 2010).

Objective Function:

$$\min \frac{1}{2} w^2 \quad \text{subject to } y_i(w \cdot x_i + b) \geq 1, \ \forall i \quad (2)$$

where

- w is the weight vector,
- b is the bias,
- y_i is the class label (+1 or −1),
- x_i is the input vector.

Figure 4 shows the SVM architecture.

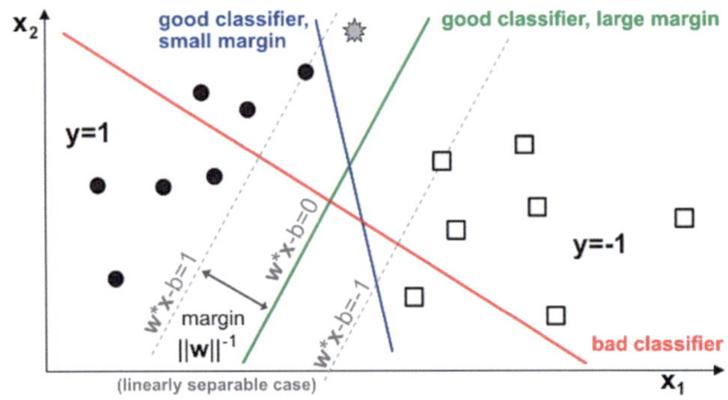

Fig. 4 The support vector machine (SVM) structure

2.3.3 eXtreme Gradient Boosting (XGB)

XGB is an ensemble learning method based on decision trees. It builds models sequentially, minimizing a loss function by applying gradient descent. XGB uses a regularized objective function to reduce overfitting.

Objective function:

$$L(\theta) = \sum_{i=1}^{n} l(y_i, \hat{y}_i) + \sum_{k=1}^{K} \Omega(f_k) \quad (3)$$

where

- l is the loss function (e.g., logarithmic loss for classification),
- $\Omega(f_k)$ is the regularization term for the complexity of the trees,
- $\hat{y}_i$ is the predicted value.

2.3.4 Random Forest (RF)

RF is an ensemble learning technique that builds multiple decision trees during training and combines their outputs for classification. Each tree is trained on a random subset of the data and features, reducing variance and overfitting (Ibrar et al. 2007).

Classification Output:

$$\hat{y} = \text{mode}(T_1(x), T_2(x), \ldots, T_n(x)) \quad (4)$$

where $T_i(x)$ represents the prediction of the i-th tree.

2.3.5 Multilayer Perceptron Regression Model (MLP)

MLP is a type of feedforward artificial neural network that learns a mapping between input features and class labels. It uses multiple layers of interconnected neurons with activation functions (Chen et al. 2020).

Forward Propagation:

$$h^{(l)} = f\left(w^{(l)}h^{(l-1)} + b^{(l)}\right) \quad (5)$$

where

- $h^{(l)}$ is the output of layer l,
- $w^{(l)}$ and $b^{(l)}$ are the weight matrix and bias vector for layer l,
- fff is the activation function (e.g., ReLU or sigmoid).

Training Objective: Minimize the cross-entropy loss for classification tasks:

$$l = -\frac{1}{n}\sum_{i=1}^{n}\sum_{k=1}^{K} y_{i,k}\log \hat{y}_{i,k} \quad (6)$$

Here $y_{i,k}$ is the true label and $\hat{y}_{i,k}$ is the predicted probability for class k. Figure 5 shows the MLP architecture.

In this study, the optimal hyperparameters for the MLP model were chosen to achieve the highest classification accuracy while ensuring a balance between model complexity and computational efficiency. These parameters are summarized in Table 5.

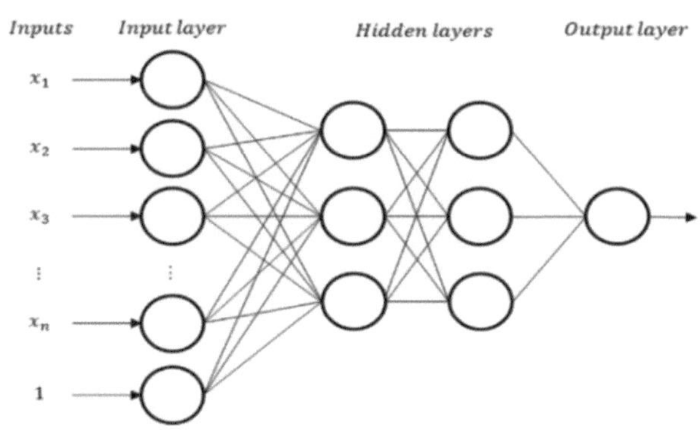

Fig. 5 The multilayer perceptron (MLP) structure

Table 5 Optimal hyperparameters of the MLP model for feedwater quality classification

Hyperparameter	Value	Description
Hidden layers	(100, 50)	Two fully connected layers with 100 and 50 neurons, respectively
Max epochs	500	Maximum number of training iterations
Activation function	ReLU (hidden layers) Softmax (output layer)	Ensures nonlinearity and multi-class classification
Optimizer	Adam	Adaptive moment estimation for efficient training
Random state	42	Ensures reproducibility of results

2.4 Metrics for Assessing Model Performance

The ML: The dataset is divided into two main parts: the training set, which consists of 80% of the data, and the test and validation set, which accounts for the remaining 20%. The test and validation data are used to fine-tune model parameters and evaluate the final model's generalization ability. During the evaluation phase, the performance of each model was assessed using widely recognized classification metrics, including accuracy, recall, precision, F1-score, and the Matthews Correlation Coefficient (MCC). These metrics provide a comprehensive assessment of the models' ability to accurately classify feedwater quality into three categories: Excellent, Good, and Poor, as illustrated in Fig. 6.

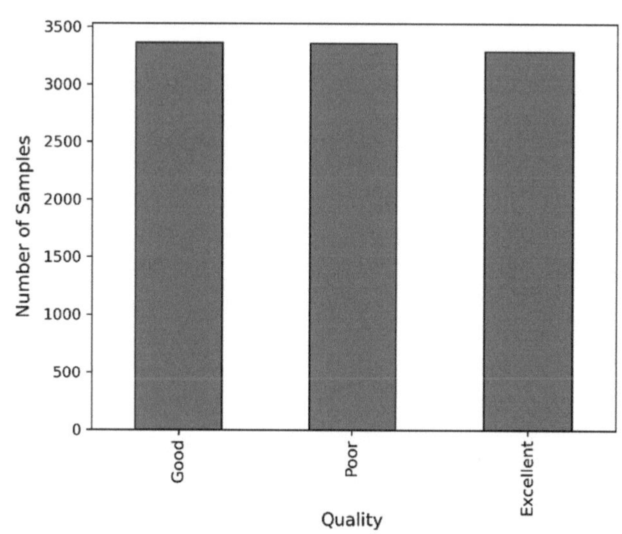

Fig. 6 Distribution of water quality classes into three categories: excellent, good, and poor

Additionally, confusion matrices were generated to offer deeper insights into the models' classification performance by visualizing misclassifications. This ensures a robust and reliable evaluation of the predictive models developed in this study.

The evaluation metrics are calculated based on the confusion matrix components:

- **True Positive (TP)**: Correctly predicted positive samples.
- **True Negative (TN)**: Correctly predicted negative samples.
- **False Positive (FP)**: Incorrectly predicted positive samples.
- **False Negative (FN)**: Incorrectly predicted negative samples.

1. **Accuracy**

Accuracy represents the overall correctness of the model, defined as the proportion of correctly classified samples:

$$\text{Accuracy} = \frac{\text{TP} + \text{TN}}{\text{TP} + \text{FP} + \text{FN} + \text{TN}} \quad (7)$$

2. **Recall (Sensitivity)**

Recall measures the model's ability to identify all positive samples:

$$\text{Recall} = \frac{\text{TP}}{\text{TP} + \text{FN}} \quad (8)$$

3. **Precision**

Precision evaluates the model's reliability in predicting positive samples:

$$\text{Precision} = \frac{\text{TP}}{\text{TP} + \text{FP}} \quad (9)$$

4. **F1-Score**

The F1-score is the harmonic mean of precision and recall, providing a balanced measure when the class distribution is imbalanced:

$$\text{F1-score} = \frac{2 \times \text{Recall} \times \text{Precision}}{\text{Recall} + \text{Precision}} \quad (10)$$

5. **Matthews Correlation Coefficient (MCC)**

The MCC evaluates the quality of binary classifications and is particularly useful for imbalanced datasets. It is a correlation coefficient between observed and predicted classifications:

$$\text{MCC} = \frac{(\text{TP} \times \text{TN}) - (\text{FP} \times \text{FN})}{\sqrt{(\text{TP} + \text{FP})(\text{TP} + \text{FN})(\text{TN} + \text{FP})(\text{TN} + \text{FN})}} \quad (11)$$

MCC values range from -1 to $+1$, where $+1$ indicates perfect prediction, 0 indicates no predictive power, and -1 indicates complete disagreement between prediction and observation.

3 Results

3.1 Performance of Classification Models

The classification models tested in this study K-NN, SVM, XGB, RF, and MLP demonstrated varying performance across different evaluation metrics such as accuracy and the Matthews Correlation Coefficient (MCC). Additionally, metrics such as recall, precision, and F1-score support the trends observed. The XGB model achieved an optimal balance between recall and precision, yielding an F1-score close to 1 across all classes. While other models, such as SVM and MLP, also performed satisfactorily, they showed a slight decline in performance for the minority classes. Table 6 presents the results, highlighting the ability of the algorithms to efficiently classify feedwater quality in reverse osmosis systems.

Table 6 Performance metrics of classification models for feedwater quality in reverse osmosis systems

Model	Accuracy (%)	Precision (%)	Recall (%)	F1-score (%)	MCC (%)
K-NN	99.23	99.00	98.00	98.00	98.52
SVM	96.56	95.00	94.00	94.00	93.40
XGB	98.8	99.00	97.50	98.00	97.68
RF	98.6	98.00	97.40	97.40	97.30
MLP	98.93	99.00	97.00	98.00	97.94

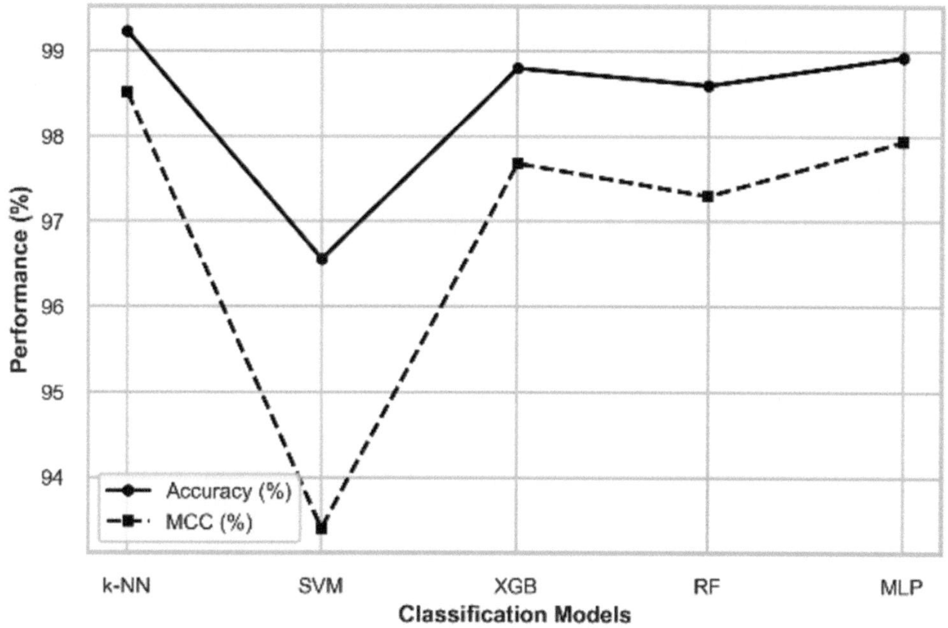

Fig. 7 Comparison of accuracy and MCC for the classification model performance

3.2 Comparison of Accuracy and MCC

The accuracy scores for the tested models range from 96.56% for the SVM model to 99.23% for the K-NN model. The K-NN model stands out for its exceptional accuracy, indicating a superior capacity to correctly predict water quality classes. Similarly, the MCC, which provides a more balanced evaluation of performance by considering both true and false predictions, shows high value for K-NN (98.52%). The other models also performed well, although slightly less well than K-NN. Figure 7 illustrates the performance of the classification models in terms of accuracy and Matthews Correlation Coefficient (MCC).

3.3 Confusion Matrix Visualization

The confusion matrices of the different models provide a detailed analysis of classification errors. For the K-NN model, misclassifications are mainly concentrated in the "Good" and "Poor" classes, with some "Good" samples incorrectly classified as "Poor." However, the majority class, "Excellent," is classified with very high accuracy. The SVM model, while effective, is more prone to errors in the "Good" class, with several samples misclassified as either "Excellent" or "Poor." On the other hand, the XGB model delivers near-perfect performance, with an exceptionally low error rate, confirming its status as the best performing model in this study. The RF model also demonstrates strong performance, slightly below XGB but offering good stability across all classes. Finally, the MLP model achieves competitive results, accurately classifying most samples, though it exhibits minor misclassification patterns similar to the RF model. Figure 8 illustrates these confusion matrices, highlighting the comparative performance of the tested models.

3.4 Impact of Classification on Operational Costs

Accurately classifying feedwater quality has a direct impact on the operational costs of desalination plants. As shown in Fig. 9, the frequency of Cleaning-In-Place (CIP) operations varies significantly depending on water quality, leading to substantial differences in chemical consumption. For "Poor" quality feedwater, CIP is required twice per month, consuming approximately 2,600 kg of cleaning chemicals per train—split equally between 1,300 kg of basic solution and 1,300 kg of acidic solution. In contrast, for "Good" quality feedwater, the CIP frequency is reduced to once or twice every 2 months, resulting in a lower chemical consumption of 1,560 kg over the same period. For "Excellent" quality feedwater, the need for CIP drops even further to once or twice every 4 months, equating to an average consumption of 325 kg per month. These findings highlight the economic benefits of maintaining high feedwater

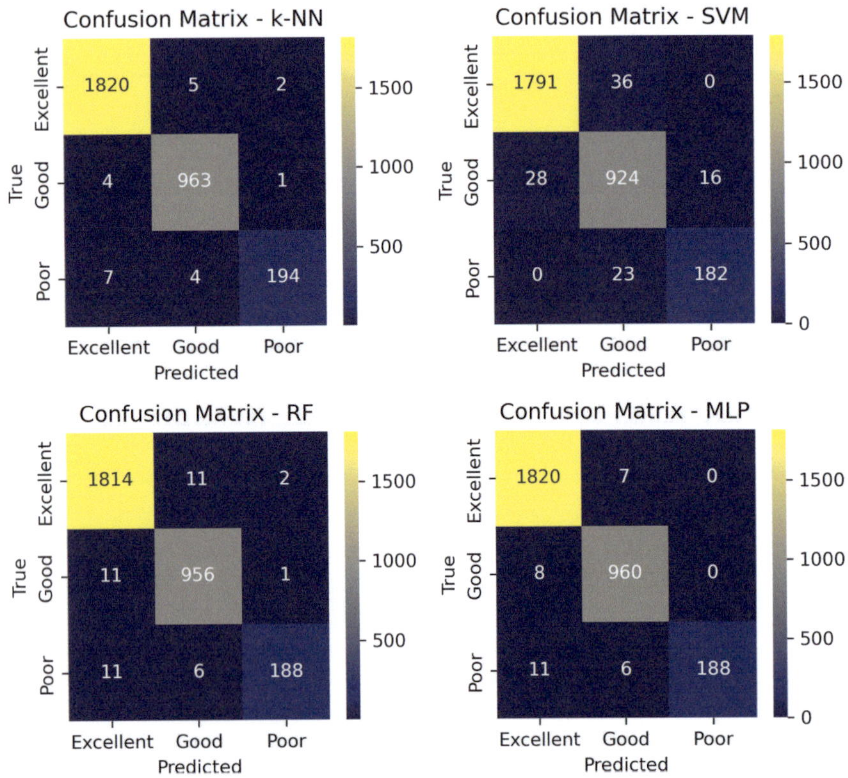

Fig. 8 Confusion matrices for the classification models, showing the true versus predicted classifications for each model

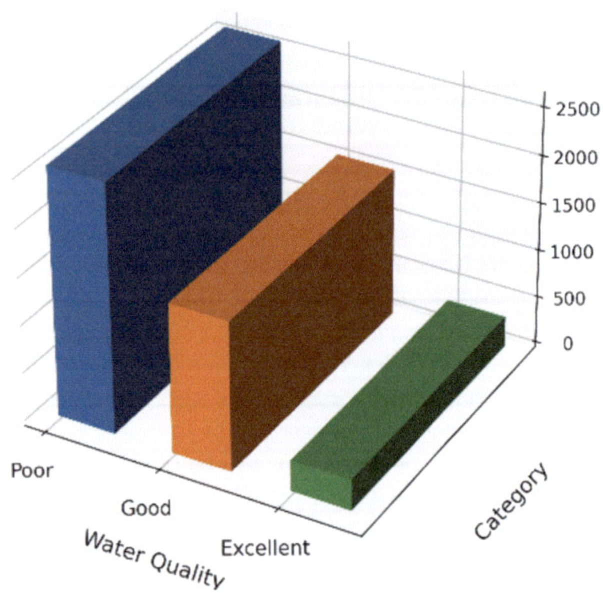

Fig. 9 Monthly cleaning product consumption (kg/month per RO train) based on feedwater quality classification

quality, as reducing CIP frequency not only lowers chemical costs but also minimizes downtime and extends membrane lifespan.

4 Discussion

The findings of this study highlight the crucial role of selecting the right classification algorithm for accurately assessing water quality in reverse osmosis (RO) systems. Among the models tested, K-NN achieved the highest accuracy (99.23%), while XGB, RF, and MLP also demonstrated strong performances with MCC scores of 97.68%, 97.30%, and 97.94%, respectively. These results align with existing literature, where XGB and RF are often praised for their ability to model complex nonlinear relationships and handle imbalanced datasets effectively. For instance, in their work, Hamada and Zaqoot (2024) found that XGB outperformed other models in predicting water quality parameters in a wastewater treatment plant, a finding reinforced by Naghibi et al. (2020) who validated the robustness of RF and XGB in groundwater source assessments. Despite its competitive performance, the MLP model could benefit from finer hyperparameter tuning, particularly for underrepresented classes. Similar studies, such as that by Wang et al. (2024), emphasize the importance of optimizing neural network architectures for surface water quality prediction, particularly when dealing with noisy or limited datasets. In contrast, SVM exhibited the lowest MCC score (93.40%), suggesting that its sensitivity to data variations may limit its effectiveness in highly dynamic water quality assessments.

When the models classify water quality as "Poor," this typically indicates the presence of contaminants, such as high dissolved solids or organic impurities, which can accelerate membrane fouling and increase operational costs. This issue is well documented in the literature, with Ibrar et al. (2007) and Alsawaftah et al. (2021) highlighting the detrimental impact of fouling on membrane efficiency and maintenance demands. By leveraging machine learning models like XGB and RF, early detection of poor water quality becomes possible, allowing operators to anticipate fouling risks and implement preventive measures. Lim et al. (2019) demonstrated how predictive models effectively identify fouling trends, reducing system downtime and optimizing membrane usage. Similarly, Shahouni et al. (2024) emphasize that advanced ML techniques like RF and XGB enhance decision-making for predictive maintenance strategies.

5 Conclusion and Perspectives

This study highlights the potential of machine learning algorithms in classifying feedwater quality for reverse osmosis systems, emphasizing the superior performance of models like K-NN, XGB, and RF. These models achieved high accuracy and MCC scores, demonstrating their ability to effectively manage data complexities and imbalances. The findings reinforce the importance of predictive maintenance in optimizing membrane performance, reducing fouling risks, and ensuring long-term operational efficiency. By enabling early detection of deteriorating water quality, these ML models help minimize maintenance costs and support sustainable desalination operations.

Looking ahead, integrating these predictive models into an IoT-based system offers a promising avenue for real-time monitoring and data-driven decision-making. By connecting sensors, data acquisition units, and cloud-based ML platforms, such a system could provide automated alerts on water quality fluctuations and membrane health, enhancing operational responsiveness. Additionally, an IoT framework would support continuous data collection, enabling further refinement of machine learning models for improved accuracy over time.

Future research should focus on developing and deploying this IoT-enabled solution, addressing challenges related to data security, scalability, and integration with existing SCADA systems. By combining predictive machine learning models with IoT technologies, the water treatment industry can move toward more efficient, resilient, and sustainable desalination processes.

Acknowledgements The authors would like to thank the laboratory teams at the MAROC CENTRAL Demineralization Station for their invaluable assistance and support.

CRediT Authors' Contribution Statement Conceptualization: [**Cherki Lahlou**]; Methodology: [**Cherki Lahlou**], Formal analysis and survey: [**Cherki Lahlou**], [**Belaid Bouikhalene**], [**Jamaa Bengourram**], [**Hassan Latrache**]; Writing—preparation of original project: [**Cherki Lahlou**]; Writing—revision and editing: [**Belaid Bouikhalene**], [**Jamaa Bengourram**], [**Hassan Latrache**], [**Radouane El Amri**]; Supervision: [**Belaid Bouikhalene**], [**Jamaa Bengourram**], [**Hassan Latrache**]; Sampling protocol: [**Radouane El Amri**].

Declaration of Competing Interest As part of the interests associated with this publication, the authors declare that they have no financial conflicts of interest or personal relationships likely to have influenced the work presented in this paper.

References

A. Abbas, "Model predictive control of a reverse osmosis desalination unit," *Desalination*, vol. 194, pp. 268–280, 2006, https://doi.org/10.1016/j.desal.2005.10.033.

A. M. Aish, H. A. Zaqoot, and S. M. Abdeljawad, "Artificial neural network approach for predicting reverse osmosis desalination plants performance in the Gaza Strip," *Desalination*, vol. 367, pp. 240–247, 2015, https://doi.org/10.1016/j.desal.2015.04.008.

S. Al, T. Bonny, S. W. Hasan, and N. Hilal, "Can machine language and artificial intelligence revolutionize process automation for water treatment and desalination ?," *Desalination*, vol. 458, no. February, pp. 84–96, 2019, https://doi.org/10.1016/j.desal.2019.02.005.

N. Alsawaftah, W. Abuwatfa, N. Darwish, and G. Husseini, "A comprehensive review on membrane fouling: Mathematical modelling, prediction, diagnosis, and mitigation," *Water*, 2021, Water 2021, https://doi.org/10.3390/w13091327.

N. Anis, S. F., Hashaikeh, R., & Hilal, "Reverse osmosis pretreatment technologies and future trends: A," *Desalination*, vol. 452, pp. 159–195, 2019.

A. Antony, J. How, S. Gray, A. E. Childress, P. Le-clech, and G. Leslie, "Scale formation and control in high pressure membrane water treatment systems: A review," *J. Memb. Sci.*, vol. 383, no. 1–2, pp. 1–16, 2011, https://doi.org/10.1016/j.memsci.2011.08.054.

A. R. Bartman, E. Lyster, R. Rallo, P. D. Christofides, and Y. Cohen, "Mineral scale monitoring for reverse osmosis desalination via real-time membrane surface image analysis," *Desalination*, vol. 273, no. 1, pp. 64–71, 2011, https://doi.org/10.1016/j.desal.2010.10.021.

J. Barzegar, R., Aalami, M. T., & Adamowski, "Short-term water quality variable prediction using a hybrid CNN – LSTM deep learning model," *Stoch. Environ. Res. Risk Assess.*, vol. 34, no. 2, pp. 415–433, 2020, https://doi.org/10.1007/s00477-020-01776-2.

L. E. O. Breiman, "Random Forests," *Mach. Learn.*, vol. 45, pp. 5–32, 2001, https://doi.org/10.1023/A:1010933404324.

X. Chen, X. Xia, P. Liang, X. Cao, H. Sun, and X. Huang, "Stacked microbial desalination cells to enhance water desalination efficiency," *Environ. Sci. Technol.*, vol. 45, no. 6, pp. 2465–2470, 2011, https://doi.org/10.1021/es103406m.

Y. Chen *et al.*, "Effect of turbidity on micropollutant removal and membrane fouling by MIEX/ultra filtration hybrid process," *Chemosphere*, vol. 216, pp. 488–498, 2019, https://doi.org/10.1016/j.chemosphere.2018.10.148.

Y. Chen, L. Song, Y. Liu, L. Yang, and D. Li, "A Review of the Artificial Neural Network Models for Water Quality Prediction," vol. 10, no. 17, p. 5776, 2020.

Drinking-water. Accessed: Dec. 06, 2023. [Online]. Available: https://www.who.int/news-room/fact-sheets/detail/drinking-water.

U. Ejaz, S. Mulk, S. Jehangir, and Z. Ahmad, "Monitoring the Industrial waste polluted stream - Integrated analytics and machine learning for water quality index assessment Monitoring the Industrial waste polluted stream - Integrated analytics and machine learning for water quality index assessment," *J. Clean. Prod.*, vol. 450, no. April, p. 141877, 2024, https://doi.org/10.1016/j.jclepro.2024.141877.

S. Fujioka, T., Ung, C. T. L., Okuda, T., & Boivin, "Controlling membrane fouling of nanofiltration using poly-aluminum chloride and Moringa oleifera coagulants," *Sep. Purif. Technol.*, vol. 4, no. 1, pp. 8–17, 1994, https://doi.org/10.1016/j.seppur.2023.126016.

S. Ghosh, "SVM - PGSL coupled approach for statistical downscaling to predict rainfall from GCM output," *J. Geophys. Res. Atmos.*, vol. 115, no. D22, 2010, https://doi.org/10.1029/2009JD013548.

L. F. Greenlee, D. F. Lawler, B. D. Freeman, B. Marrot, P. Moulin, and P. Ce, "Reverse osmosis desalination: Water sources, technology, and today's challenges," *Water Res.*, vol. 43, no. 9, pp. 2317–2348, 2009, https://doi.org/10.1016/j.watres.2009.03.010.

H. Gu, A. R. Bartman, M. Uchymiak, P. D. Christofides, and Y. Cohen, "Self-adaptive feed flow reversal operation of reverse osmosis desalination," *Desalination*, vol. 308, pp. 63–72, Jan. 2013, https://doi.org/10.1016/J.DESAL.2012.07.041.

M. S. Hamada and A. Zaqoot, "Using a supervised machine learning approach to predict water quality at the Gaza wastewater treatment plant," pp. 132–144, 2024, https://doi.org/10.1039/d3va00170a.

Hibbs MR, McGrath, Lucas K, Kang S, Adout A, Altman S J, Elimelech M, Cornelius C J (2016), "Designing a biocidal reverse osmosis membrane coating: Synthesis and biofouling properties," *Desalination*, vol. 380, pp. 52–59, 2016, https://doi.org/10.1016/j.desal.2015.11.017.

A. Ibrar, I., Naji, O., Sharif, A., Malekizadeh, A., Alhawari, A., Alanezi, A. A., & Altaee, "A Review of Fouling Mechanisms, Control Strategies and Real-Time Fouling Monitoring Techniques in Forward Osmosis," *Water*, 2007, https://doi.org/10.3390/w11040695.

S. J. Im, N. D. Viet, and A. Jang, "Real-time monitoring of forward osmosis membrane fouling in wastewater reuse process performed with a deep learning model," *Chemosphere*, vol. 275, p. 130047, 2021, https://doi.org/10.1016/j.chemosphere.2021.130047.

S. Ji, Y. Mi, H. Park, S. Ki, and K. Jeong, "Enhancing accuracy of membrane fouling prediction using hybrid machine learning models," *Desalin. Water Treat.*, vol. 146, no. 2019, pp. 22–28, https://doi.org/10.5004/dwt.2019.23444.

Y. Jin, H. Lee, C. Park, and S. Hong, "ASTM Standard Modified Fouling Index for Seawater Reverse Osmosis Desalination Process: Status, Limitations, and Perspectives," *Sep. Purif. Rev.*, vol. 49, no. 1, pp. 55–67, 2020, https://doi.org/10.1080/15422119.2018.1515777.

M. T. Khan, M. Busch, V. G. Molina, A. H. Emwas, C. Aubry, and J. P. Croue, "How different is the composition of the fouling layer of wastewater reuse and seawater desalination RO membranes?," *Water Res.*, vol. 59, pp. 271–282, 2014, https://doi.org/10.1016/j.watres.2014.04.020.

J. A. Kubat, M., & Kubat, *An Introduction to Machine Learning*, 3rd ed. Cham, Switzerland: Springer International Publishing, 2017. https://doi.org/10.1007/978-3-319-63913-0.

Y. Lebrini, A. Boudhar, R. H. H. Lionboui, and L. E. R. Arrach, "Identifying Agricultural Systems Using SVM Classification Approach Based on Phenological Metrics in a Semi - arid Region of Morocco," *Earth Syst. Environ.*, vol. 3, no. 2, pp. 277–288, 2019, https://doi.org/10.1007/s41748-019-00106-z.

S. S. Madaeni, N. Ghaemi, and H. Rajabi, "1 - Advances in polymeric membranes for water treatment," in *Advances in Membrane Technologies for Water Treatment*, Elsevier Ltd, 2015, pp. 3–41. https://doi.org/10.1016/B978-1-78242-121-4.00001-0.

Mancosu N, Snyder RL, Kyriakakis G, Spano D. (2015) Mancosu N, Snyder, R L, Kyriakakis G, "Water scarcity and future challenges for food production," *Water (Switzerland)*, vol. 7, no. 3, pp. 975–992, 2015, https://doi.org/10.3390/w7030975.

G. Meyers, Z. Kapelan, and E. Keedwell, "Short-term forecasting of turbidity in trunk main networks," *Water Res.*, vol. 124, pp. 67–76, 2017, https://doi.org/10.1016/j.watres.2017.07.035.

S. A. Naghibi, H. Hashemi, R. Berndtsson, and S. Lee, "Application of extreme gradient boosting and parallel random forest algorithms for assessing groundwater spring potential using DEM-derived factors," *J. Hydrol.*, 2020, https://doi.org/10.1016/j.jhydrol.2020.125197.

J. A. Oshiro, T. M., Perez, P. S., & Baranauskas, "How Many Trees in a Random Forest ?," in *Machine Learning and Data Mining in Pattern Recognition: 8th International Conference, MLDM 2012, Berlin, Germany, Proceedings*, 2012, pp. 154–168. https://doi.org/10.1007/978-3-642-31537-4.

N. Sadek, M. Attia, and I. El-gafy, "Experimental performance analysis for reverse osmosis pilot plant subjected to different brackish salinity spectrum," *J. King Saud Univ. - Eng. Sci.*, no. xxxx, 2023, https://doi.org/10.1016/j.jksues.2023.09.002.

Santoro C, Abad FB, Serov A, Kodali M, Howe KJ, Soavi F, Atanassov P (2017), "Supercapacitive microbial desalination cells: New class of power generating devices for reduction of salinity content," *Appl. Energy*, vol. 208, no. October, pp. 25–36, 2017, https://doi.org/10.1016/j.apenergy.2017.10.056.

R. Shahouni, M. Abbasi, M. Dibaj, and M. Akrami, "Utilising Artificial Intelligence to Predict Membrane Behaviour in Water Purification and Desalination," *Water*, 2024, https://doi.org/10.3390/w16202940.

M. A. Shannon, P. W. Bohn, M. Elimelech, J. G. Georgiadis, B. J. Mariñas, and A. M. Mayes, "Science and technology for water purification in the coming decades," *Nature*, vol. 452, no. 7185, pp. 301–310, 2008, https://doi.org/10.1038/nature06599.

A. G. Sim, L. N., Chong, T. H., Taheri, A. H., Sim, S. T. V., Lai, L., Krantz, W. B., & Fane and A. G. Fane, "A review of fouling indices and monitoring techniques for reverse osmosis," *Desalination*, vol. 434, pp. 169–188, 2018, https://doi.org/10.1016/j.desal.2017.12.009.

G. Singh, K. P., Basant, A., Malik, A., & Jain, "Artificial neural network modeling of the river water quality — A case study," *Ecol. Modell.*, vol. 220, no. 6, pp. 888–895, 2009, https://doi.org/10.1016/j.ecolmodel.2009.01.004.

X. Tang *et al.*, "Respective role of iron and manganese in direct ultrafiltration : from membrane fouling to flux improvements," *Sep. Purif. Technol.*, vol. 259, no. September 2020, p. 118174, 2021, https://doi.org/10.1016/j.seppur.2020.118174.

X. Wang, Y. Dong, J. Yang, Z. Liu, and J. Lu, "A benchmark-based method for evaluating hyperparameter optimization techniques of neural networks for surface water quality prediction," *Front. Environ. Sci. Eng.*, 2024, https://doi.org/10.1007/s11783-024-1814-5.

D. P. Webb, G. Skouteris, and S. Rahimifard, "In-plant real-time manufacturing water content characterisation," *Water Resour. Ind.*, vol. 20, pp. 37–45, Dec. 2018, https://doi.org/10.1016/J.WRI.2018.08.003.

A Bioinformatics and Experimental Characterization of Extracellular Esterase from *Lactiplantibacillus plantarum* FSO1 and *Candida pelliculosa* L18 Isolated from Traditional Fermented Green Olive

Yahya Rokni, Reda Bellaouchi, Houssam Abouloifa, Ismail Hasnaoui, Nabil Ghabbour, Chorouk Zanane, Sara Gaamouche, Kaoutar Mchiouer, Aziz Hirri, and Abdeslam Asehraou

Abstract

Table olives are one of the most important traditional fermented foods in the Mediterranean Basin. However, the presence of phenolic compounds in fresh olives, particularly the bitter glucoside oleuropein, necessitates the conversion of this compound into non-bitter substances to make the fruit edible. Industrial table olive processing typically involves chemical treatments, which pose environmental risks and potential hazards for both the industry and consumers. Therefore, the development of an innovative biological treatment process for green table olive production could significantly enhance the Moroccan olive industry. This study investigates esterase activity of *Lactiplantibacillus plantarum* FSO1 and *Candida pelliculosa* L18, both isolated from green olives during fermentation, as potential bio-debittering agents. The activity was maximum at pH 6 and 0% NaCl, with enhanced activity of *C. pelliculosa* L18 over *L. plantarum* FSO1. Bioinformatics and molecular research confirmed the presence of esterase genes, which indicate their potential as clean alternatives to chemical debittering agents employed in olive processing. These outcomes provide the basis for industrial exploitation of sustainable olive fermentation. The results indicate that lactic acid bacteria (LAB) and yeast are both capable of producing extracellular, inducible esterase enzymes of potential application. These enzymes possess high value for the biological treatment of table olives as a green alternative to conventional chemical processing pathways.

Keywords

Oleuropein · Esterase · Fermented green olives · *Lactiplantibacillus plantarum* · *Candida pelliculosa*

Y. Rokni (✉) · C. Zanane
Research Unit Bioprocess and Biointerfaces, Laboratory of Industrial Engineering and Surface Engineering, FST, Sultan Moulay Slimane University, BP 523, Beni Mellal 23000, Morocco
e-mail: y.rokni@usms.ma

Y. Rokni · R. Bellaouchi · H. Abouloifa · I. Hasnaoui · N. Ghabbour · S. Gaamouche · A. Asehraou
Laboratory of Bioresources, Biotechnologies, Ethnopharmacology, and Health, Faculty of Sciences, University Mohamed Premier, BP 717, Oujda 60000, Morocco

H. Abouloifa
Research Unit of Microbiology, Biomolecules and Biotechnology, Laboratory of Chemistry Physics and Biotechnology of Molecules and Materials, Faculty of Sciences and Techniques – Mohammedia, Hassan II University of Casablanca, BP 146, Mohammedia 28806, Morocco

N. Ghabbour
Laboratory of Natural Resources and Environment, Polydisciplinary Faculty of Taza, Sidi Mohamed Ben Abdellah University, BP 1223, Taza 35000, Morocco

C. Zanane
Higher Institutes of Nursing and Health Technical Professions Béni Mellal (ISPITS), Beni Mellal 23000, Morocco

K. Mchiouer
Higher Institutes of Nursing and Health Technical Professions Nador (ISPITS), Nador 62000, Morocco

A. Hirri
National School of Applied Sciences, Sultan Moulay Slimane University, Beni Mellal 23000, Morocco

M. Gouskir et al. (eds.), *Cutting-Edge Artificial Intelligence Applications*, Sustainable Artificial Intelligence-Powered Applications,
https://doi.org/10.1007/978-3-032-00125-2_12

1 Introduction

Oleuropein, the main olive polyphenol glucoside, is the bitter principle that is accountable for olive bitterness (Gikas et al. 2006; Charoenprasert and Mitchell 2012). The table olive industry has conventionally used lye (NaOH) treatment to debitter fruits, chemical debittering of olives using NaOH poses environmental and health risks, creating a need for sustainable biological alternatives (Rokni et al. 2017). Being an eco-friendly alternative, biological debittering employs microbial starter cultures that produce enzymes such as esterases and β-glucosidase (Rokni et al. 2021). The enzymes hydrolyze oleuropein and other bitter compounds into non-bitter compounds, enabling eco-friendly valorization of olives.

Enzymatic treatment includes two important steps: oleuropein is hydrolyzed to glucose and oleuropein aglycone by β-glucosidase, and the aglycone is further hydrolyzed by esterases to health-promoting compounds such as hydroxytyrosol a very active antioxidant present in high quantity in olive oil and elenolic acid (Fayek et al. 2021; Vaccalluzzo et al. 2022). Hydrolysis not only debitters but also improves fermentation by inducing lactic acid production, of special interest to olive preservation and for nutritional, technological, and health-related property improvement (Anagnostopoulos et al. 2020). These microorganisms like *Lactiplantibacillus plantarum* (lactic acid bacteria) and Candida species (yeast) produce these enzymes, and therefore their genetic regulation, production, and activity must be elucidated for industrial application.

This study combines bioinformatics and experimental approaches to characterize esterase enzymes from *L. plantarum* FSO1 and *C. pelliculosa* L18, offering a sustainable solution for olive debittering. Through the use of bioinformatics databases and molecular biology techniques such as 16S rDNA sequencing, we identified the strains and examined the genetic basis of the production of the enzymes. From the bacterial genomic sequences, we plan to search for esterase-coding genes, create enzyme optimization for activity, and determine patterns of expression to further drive industrial fermentation.

It combines classical microbiology and bioinformatics, and the research elucidates the molecular mechanisms of enzyme production. The knowledge created will improve fermentation efficiency, enhance debittering processes, and encourage the quality of fermented olive products, bridging biotechnology with sustainable food processing.

The aim of this study is to characterize the esterase activity of *L. plantarum* FSO1 and *C. pelliculosa* L18, optimize enzyme productivity, and investigate their potential for industrial applications.

2 Material and Methods

2.1 Microorganisms and Culture Conditions

The microbial strains used in this study are *L. plantarum* FSO1 and *C. pelliculosa* L18. These strains have been isolated from brines of green olives which were in natural fermentation (Rokni et al. 2015), and selected for their oleuropenolytic activity (Rokni 2017).

2.2 Enzyme Activity Assays

(a) **API-ZYM System**

The enzymatic characterization of strains was performed using the API-ZYM kit (BioMérieux, France). This system was used to determine the enzymatic profile of the oleuropein-degrading strains (FSO1 and L18) (Alfano et al. 2011), following the manufacturer's instructions. Microbial suspensions were dispensed into the API-ZYM™ cupules and incubated at 37 °C for 4 h. After incubation, ZYM A and ZYM B reagents were added, and color changes were observed for approximately 10 s under a light source (Principi et al. 2003; Ahmadova et al. 2013).

(b) **Localization of Esterase Activity**

The strains studied in this work (*L. plantarum* FSO1 and *C. pelliculosa* L18) were tested to determine the localization of their esterase activity (extracellular, intracellular, or membrane-bound). For this purpose, overnight cultures (5 μL) of the strains were inoculated into 5 mL of standard MRS medium and modified MRS medium. The modified MRS contained 1% (w/v) oleuropein (Extrasynthese, Genay, France) as the sole carbon source and was incubated at 30 °C for 7 days. The cultures were centrifuged at 12,000×*g* for 12 min at 4 °C (Hermle Labnet Z216MK). The supernatant was carefully separated from the cell pellet. The cell pellet was resuspended in 0.1 M phosphate buffer (pH 5), and the cells were lysed by sonication at 4 °C in an ultrasonic bath (5 s pulses with 5 s intervals for 45 min; Bandelin Sonorex Digitec). The lysed cells were centrifuged again at 12,000×*g* for 12 min at 4 °C to separate the intracellular and membrane fractions. Following these steps, three fractions were obtained for each strain to measure enzymatic activity: the extracellular fraction (supernatant 1: S1), the intracellular fraction (supernatant 2: S2), and the membrane fraction (M).

(c) **SDS Gel Electrophoresis**

Proteins from microbial cultures grown with oleuropein (1%) as the sole carbon source were precipitated using $(NH_4)_2SO_4$ (80% saturation). After centrifugation, the pellets were dissolved in sodium acetate buffer (pH 5.0). Proteins were separated by molecular weight using SDS-PAGE under reducing conditions, following the Laemmli (1970) method. A 15% separating gel and 5% stacking gel containing 0.1% SDS were used, with Tris–glycine buffer (pH 8.3) as the electrode buffer. Samples were boiled with sample buffer before loading, and electrophoresis was performed at 130 V for 40 min. Molecular weights were determined using protein standards, and proteins were visualized by silver nitrate staining (Wray et al. 1981).

(d) **Determination of Esterase Activity**

Esterase activity was determined using the method of Mackness et al. (1983) on enzyme extracts (S1, S2 supernatants and membrane fraction (M)). The reaction mixture contained 50 μL enzyme extract, 50 μL 150 mM paranitrophenyl acetate (p-NPA) (Merck) in ethanol and 2.9 mL Tris–HCl buffer (9.2 mM, pH 7.5) to give a final concentration of 2.5 mM p-NPA in a total volume of 3 mL. The reaction was initiated by the addition of enzyme extract (S1, S2 supernatants and membrane fraction (M)) and incubated at 25 °C for 4 min.

The amount of paranitrophenol (p-NP) released was determined spectrophotometrically by measuring the absorbance of the solution at 405 nm. A unit of enzyme activity (U) was defined as the amount of enzyme that produced 1 μmol of p-NP per minute under the conditions of the experiment. A calibration curve was prepared using p-NP.

(e) **Optimization of Esterase Production**

The oleuropeinolytic strains studied *L. plantarum* (FSO1), *C. pelliculosa* (L18), and their combination (FSO1/L18) were grown in MRSm broth (5 mL) supplemented with oleuropein (1%) as the sole carbon source. After 7 days of incubation at 30 °C, supernatants (S1) from each culture were carefully collected to measure esterase enzyme activity.

Esterase production was monitored in vitro by measuring the rate of hydrolysis of p-nitrophenyl acetate (p-NPA) by esterases present in the extracts (S1 supernatants) from lactic acid bacteria (FSO1), yeast (L18), and their combination (FSO1/L18). This hydrolysis reaction produces p-nitrophenol (p-NP), a colored compound that can be quantified by measuring absorbance at 405 nm. The variation in the quantity of the product was measured over a 9-day period using the method described above.

In addition to monitoring esterase production, pH was measured using a pH meter, and the inoculated strains were enumerated for both pure and mixed cultures.

- **Effect of pH on Esterase Production**

The effect of three pH values (4, 5, and 6) on esterase production was evaluated in the different strains (*L. plantarum* FSO1, *C. pelliculosa* L18, and their combinations (FSO1/L18)). Extracts from each culture (S1 supernatants) were prepared, and esterase activity was measured using the same protocol described above, with p-NPA as the substrate. Two controls were included in this experiment: the first containing Tris–HCl buffer (9.2 mM, pH 7.5) and the second containing pure esterase. All tests were performed in triplicate. The MRSm medium was acidified to pH 6, 5, and 4 using 2 N HCl.

- **Effect of NaCl on Esterase Production**

Esterase production was assayed in the presence of varying concentrations of NaCl (0, 2, 4, 6, 8, and 10%) using 1% (v/v) inocula of *L. plantarum* (FSO1), *C. pelliculosa* (L18), and their combination (FSO1/L18). The cultures were incubated at 30 °C for 7 days. Esterase activity in the S1 supernatants from the selected oleuropeinolytic strains was determined using the same protocol described above, with p-NPA as the substrate. Two controls were included in this experiment: the first containing Tris–HCl buffer (9.2 mM, pH 7.5) and the second containing pure esterase. All assays were performed in triplicate.

- **Effect of pH and NaCl on Esterase Production**

The study of esterase production allowed us to identify the optimal combination of pH and NaCl concentration in the presence of oleuropein (1%) for maximum esterase production in the extracts of each culture studied (S1 supernatants). The selected pH was the one that yielded the highest activity for each strain: pH 6 for *L. plantarum* FSO1, *C. pelliculosa* L18, and their combination (FSO1/L18). Similarly, the NaCl concentration chosen was the one that resulted in the highest esterase activity for the studied strains (0 and 2% NaCl for *L. plantarum* FSO1, *C. pelliculosa* L18, and the FSO1/L18 combination).

All culture assays in MRSm medium were inoculated with 1% (v/v) of young cultures of the studied strains and incubated at 30 °C for 7 days. The β-glucosidase and esterase activities in the extracts from each culture (S1 supernatants) were measured using the same protocol described above.

2.3 Molecular Identification of Strains

(a) **DNA Extraction**

DNA was extracted from biochemically identified strains of lactic acid bacteria (*L. plantarum* FSO1) using the Isolate II Genomic DNA Kit (Bioline, a Meridian Life Science company, Germany). The strains were cultured in liquid MRS medium and incubated for 24 h at 30 °C. The isolated DNA was quantified and stored at −20 °C for further analysis.

(b) **DNA Quantification**

DNA quantification was performed spectrophotometrically using a NanoDrop 8000 spectrophotometer at 260 nm (Thermo Fisher Scientific Inc., MA, USA). Before starting the measurements, a blank is made with milliQ water. All measurements are recorded in real time on a computer. Nucleic acid absorbs at 260 nm, proteins at 280 nm, and organic compounds such as phenol at 230 nm. Thus, the 260 nm/280 nm and 260 nm/230 nm ratios provide information on sample purity. In theory, a ratio below 1.8 and 2.2, respectively, is accepted as pure. The DNA fractions obtained were diluted to a concentration of 150 ng/μL for the PCR reaction.

(c) **PCR Protocol**

Extracted DNA was amplified by PCR (polymerase chain reaction) using specific primers FD1 (5′-AGTTTGATCCTGGCTCAG-3′) and RP2 (5′-AAGGAGGTGATCCAGCC-3′) for the molecular identification of the bacterial 16S rDNA gene (Weisburg et al. 1991). The PCR was carried out using the MyTaq DNA Polymerase Kit (Bioline).

(d) **Purification of PCR Products**

PCR products were purified using the ExoSAP-IT PCR Product Cleanup Kit (Affymetrix) to remove residual salts, unused primers, and free nucleotides (dNTPs) from the PCR reaction. For the ExoSAP-IT treatment, 537 μL of PCR product was mixed with 2 μL of ExoSAP-IT, followed by incubation at 37 °C for 15 min in a thermocycler to degrade primers and free nucleotides. This was followed by a second incubation at 80 °C for 15 min to inactivate the ExoSAP-IT enzyme.

Once purified, the PCR products were ready for sequencing.

Table 1 Primers used for the detection of esterase genes in lactic acid bacterial strains

Primers	Primer sequence (5′-3′)	Application	Ref.
Est-1	GCTAATTTGTAACCGTATCCGCC-fwd	Esterase gene	Mtshali et al. (2010)
Est-2	CGCGCATGTTAACTTTTAGTAGAAC-rev	Esterase gene	

2.4 Bioinformatics Analysis

For each purified PCR product, a bidirectional sequencing reaction was performed. The sequencing reaction was carried out using the BigDye Terminator Cycle Sequencing Kit v3.1 (Applied Biosystems) following the manufacturer's instructions.

Sequence Analysis

The obtained sequences were analyzed using Sequencing Analysis v5.3.1 software (Applied Biosystems) and subsequently compared with sequences available in the GenBank database using the BLAST tool from the National Center for Biotechnology Information (NCBI) (http://www.ncbi.nlm.nih.gov/blast/Blast.cgi). The comparison of the sequences with the database was performed for bacterial identification. The identification criteria, as suggested by several studies (Harris et al. 2002; Clarridge 2004), were based on sequence identity thresholds: 97% identity in the 16S rRNA gene sequence for genus-level identification and 99% identity for species-level identification.

(a) **Molecular Detection of Esterase Genes**

To detect the presence of esterase genes from table olive Lactobacillus species, isolates were screened using PCR. The bacterial isolate was first cultured on normal MRS broth at 30 °C for 24 h.

DNA extraction was performed using Bioline's Isolate II genomic DNA kit, followed by DNA quantification using the NanoDrop 8000 at 260 nm. Isolated DNA was quantified and stored at −20 °C.

(b) **PCR Amplification of Esterase Genes**

DNA extracts were diluted to a concentration of 150 ng/μL for use in PCR. A set of primers specific to the esterase gene (Mtshali et al. 2010) was used (Table 1).

The sequencing reaction protocol described above was applied to the esterase PCR products following purification with the ExoSAP-IT enzyme (Affymetrix).

3 Results and Discussion

3.1 Enzyme Profile of Oleuropeinolytic Strains

(a) **Semi-quantitative API-ZYM System**

Semi-quantitative API-ZYM test was used to detect the presence and initial quantification of esterase activity in *L. plantarum* FSO1 and *C. pelliculosa* L18. The results indicated that the two strains (FSO1 and L18) produce a range of various enzymes with different qualitative and quantitative profiles (Table 2). Among the various enzymes, esterase (C4) and esterase lipase (C8) were detected in the two strains. In the API-ZYM system, esterase lipase (C8) was 1 (5 nmol) for both strains, and esterase (C4) was 1 (5 nmol) for FSO1 and 2 (10 nmol) for L18. The other enzymes of nutrient catabolism detected at low levels were naphthol phosphohydrolase, β-galactosidase, alkaline and acid phosphatases, lipase (C14), leucine arylamidase, valine arylamidase, cystine arylamidase, trypsin, acid phosphatase, *N*-acetyl-β-glucosaminidase, α-mannosidase, and α-fucosidase.

The presence of esterase and esterase lipase (C4 and C8) in both strains suggests that they have the ability to hydrolyze a broad range of substrates, making them eligible for various industrial applications.

(b) **Induction and Localization of Esterase**

Induction and localization of the esterase activity in the tested strains are shown in Fig. 1. In oleuropein as sole carbon source, the values of esterase activity detected in the extracellular fraction were 0.63 U/mL for *C. pelliculosa* L18 and 0.09 U/mL for *L. plantarum* FSO1. Conversely, in the absence of oleuropein, esterase activity was very low in all fractions (extracellular: E, intracellular: I, and cell wall: M) of both strains. This result is in agreement with that found by Bonatsou et al. (2015).

(c) **SDS-PAGE**

The production of esterase by the studied microbial strains was confirmed under denaturing conditions using SDS-PAGE with silver nitrate staining, revealing a distinct band with a molecular weight of approximately 28 kDa (Fig. 2). The electrophoretic profile of *L. plantarum* FSO1 confirmed the presence of an esterase band at 28 kDa (lane 1, Fig. 2). Similarly, the yeast *C. pelliculosa* L18 also exhibited a 28 kDa band corresponding to esterase (lane 2, Fig. 2).

Table 2 Enzymatic activity of oleuropeinolytic strains obtained by API-ZYM system

Enzymes	Enzymatic activity (nmol/5 h)	
	L. plantarum FSO1	*C. pelliculosa* L18
Control (no substrate)	0	0
Phosphatase alkaline	2	2
Esterase (C4)	1	2
Esterase lipase (C8)	1	1
Lipase (C14)	1	1
Leucine arylamidase	1	0
Valine arylamidase	1	2
Cystine arylamidase	1	2
Trypsine	1	2
α-Chymotrypsine	0	1
Phosphatase acid	2	2
Naphthol phosphohydrolase	3	2
α-Galactosidase	0	0
β-Galactosidase	1	2
β-Glucuronidase	0	0
α-Glucosidase	4	4
β-Glucosidase	2	4
N-Acetyl-β-glucosaminidase	2	0
α-Mannosidase	0	1
α-Fucosidase	0	1

The scale of the API-ZYM test was used for enzymatic quantification, 0 = no enzyme, 1 = 5 nmol, 2 = 10 nmol, 3 = 20 nmol, 4 = 30 nmol and 5 = 40 nmol or more

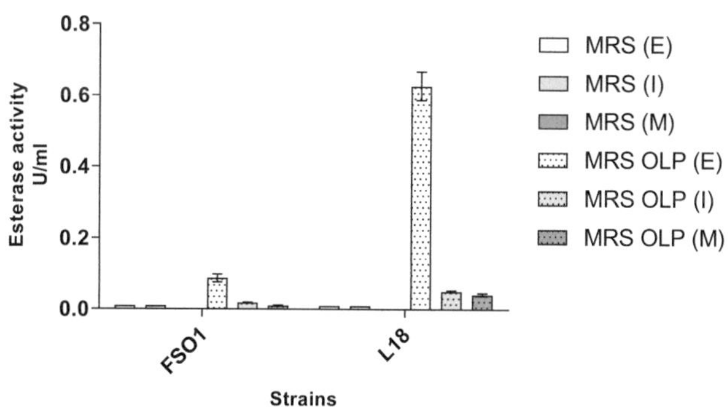

Fig. 1 Esterase activity of *L. plantarum* FSO1 and *C. pelliculosa* L18 obtained after 7 days of culture at 30 °C on MRS and modified MRS (MRSm) containing oleuropein (1%) as a sole carbon source. *Legend:* Esterase activity in (E): extracellular; (I): intracellular and (M): parietal fractions

(d) **Production of Esterase**

Esterase activity monitoring is shown in Fig. 3. *L. plantarum* FSO1 exerted very weak esterase activity and did not exceed 0.11 U/mL on the sixth day of incubation and decreased

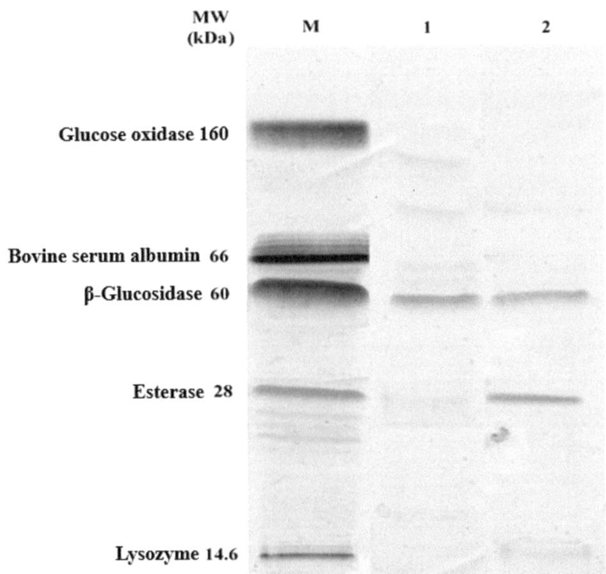

Fig. 2 Electrophoretic profile (SDS-PAGE) of proteins extracts of *L. plantarum* FSO 1 (lane 1) and *C. pelliculosa* L18 (lane 2) cultures obtained in modified MRS broth (containing OLP 1% as sole carbon source) after 7 days of culture at 30 °C (M: molecular weight (kDa) of standards)

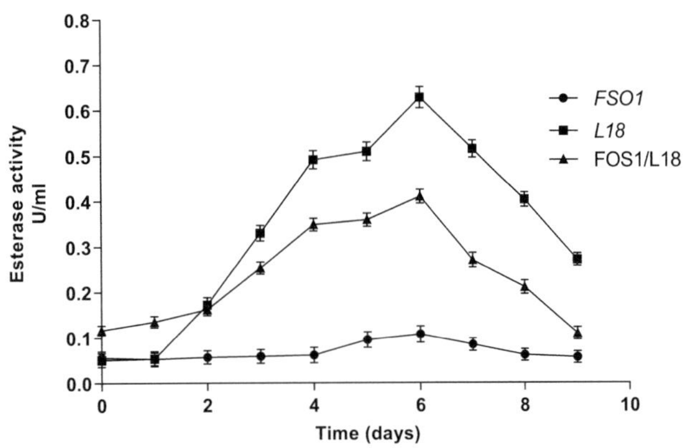

Fig. 3 Monitoring of esterase production in the oleuropeinolytic strains studied. *Legend* FSO1: *L. plantarum*, L18: *C. pelliculosa*

afterward. Conversely, *C. pelliculosa* L18 yeast exhibited much stronger esterase activity compared to the lactic acid bacterium (FSO1) with a peak of 0.63 U/mL on the sixth day of incubation, which decreased afterward.

Esterase activity profile of the mixed culture of *L. plantarum* and *C. pelliculosa* was intermediary in nature: it was lower than the pure culture of *C. pelliculosa* L18 but higher than the *L. plantarum* FSO1 culture and attained a maximum value of 0.41 U/mL on the sixth day of incubation.

Lastly, esterase activity was significantly greater in yeasts to be studied compared to lactic acid bacteria. Table olive fermentation, however, remains dependent on lactic acid bacteria. In addition, co-culture of yeast and lactic acid

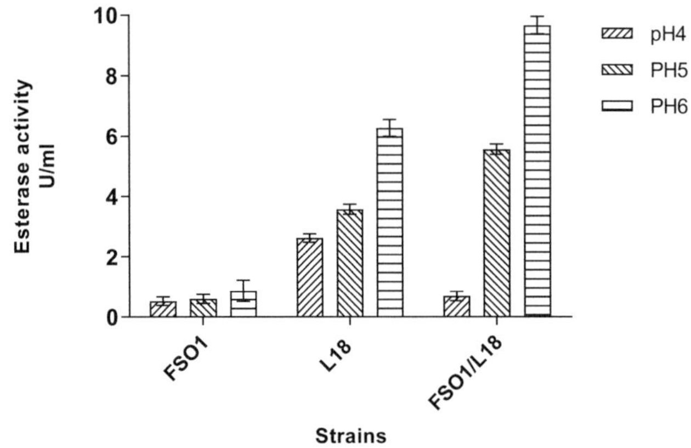

Fig. 4 Esterase activity of *L. plantarum* FSO1, *C. pelliculosa* L18, *C.* and *L. plantarum* FSO1/*C. pelliculosa* L18, examined in MRS broth with oleuropein (1%) as sole carbon source at different pHs

bacteria yielded greater esterase activity compared to *L. plantarum* pure culture.

(e) Optimization of Esterase Production

Cultures were conducted at various pHs to determine the optimum pH for the maximum production of esterase. The optimum pH was 6.0 for all the examined strains, including *L. plantarum* FSO1 and *C. pelliculosa* L18, *L. plantarum* FSO1/*C. pelliculosa*. Esterase production was remarkably lower in *L. plantarum* FSO1 compared to the other strains (Fig. 4).

These results highlight the striking influence of pH on esterase production among the different inocula used.

The effect of NaCl (2–10%) on the production of esterase by the tested strains is illustrated in Fig. 5. The maximum enzymatic activities observed in the absence of NaCl (0%) were 0.98, 7.86, and 8.08 U/mL for *L. plantarum* FSO1, *C. pelliculosa* L18, and the mixture of *L. plantarum* FSO1/*C. pelliculosa* L18, respectively. As can be observed, there were no striking differences among 0%, 2%, 4%, and 6% (w/v) NaCl concentrations, i.e., NaCl did not have any striking inhibitory effect on the esterase activity of any of the strains. Indeed, for all strains except *L. plantarum* FSO1, esterase activity was practically unaffected at NaCl concentrations of up to 6% (w/v) (Fig. 5). These findings are consistent with those of Soliman et al. (2007).

NaCl concentration is crucial in olive fermentation preservation because it inhibits the activity of undesirable microorganisms. The impact of NaCl on the hydrolysis of esters by the strains was thus studied.

Combined action of pH and NaCl for esterase production is shown in Fig. 6. The *C. pelliculosa* L18 and *L. plantarum* FSO1/*C. pelliculosa* L18 strains under investigation exhibited enhanced esterase activity without (0%) and with the addition of 2% NaCl at pH6. In pure cultures, maximum activity was observed in the *C. pelliculosa* L18

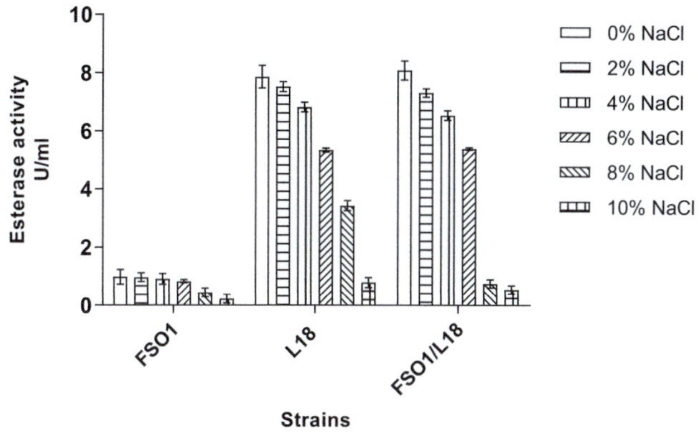

Fig. 5 Effect of NaCl concentration on esterase production on modified MRS medium (pH 6.8) after 7 days incubation at 30 °C. *Legend L. plantarum* FSO1, *C. pelliculosa* L18, combination *L. plantarum* FSO1/ C. *pelliculosa* L18

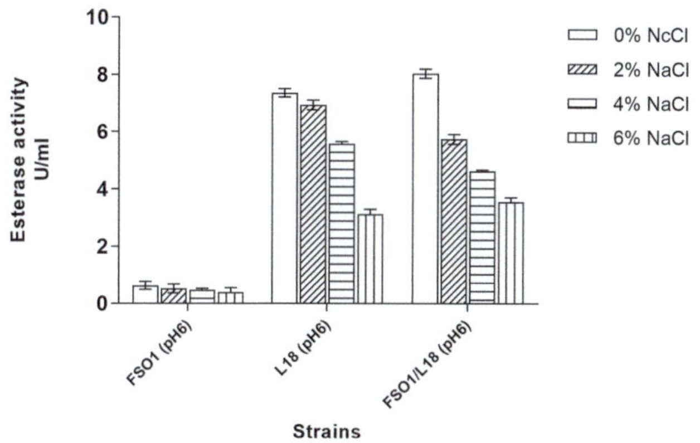

Fig. 6 Effect of pH and NaCl blend on esterase activity on supplemented MRS medium after 7 days incubation at 30 °C. *Legend L. plantarum* FSO1, *C. pelliculosa* L18, the combination *L. plantarum* FSO1/*C. pelliculosa* L18

strain, followed by the *L. plantarum* FSO1/*C. pelliculosa* L18 co-culture. We noted that with the combination studied at pH6, *L. plantarum* FSO1/*C. pelliculosa* L18, the esterase activity did not undergo a notable change compared to the pure strains' instances.

Thus pH 6 together with NaCl (0 and 2%) positively affects the production of esterase in yeast, as their combination with lactic acid bacteria does, unlike natural MRSm medium pH (6.8) (Fig. 6). Esterase activity is low with the various combinations in *L. plantarum* FSO1, though.

The optimal pH (6) and NaCl concentration (0%) for esterase production align with typical olive fermentation conditions, making these enzymes suitable for industrial applications.

This study was on a small scale; future studies should consider large-scale production and application in industrial olive fermentation.

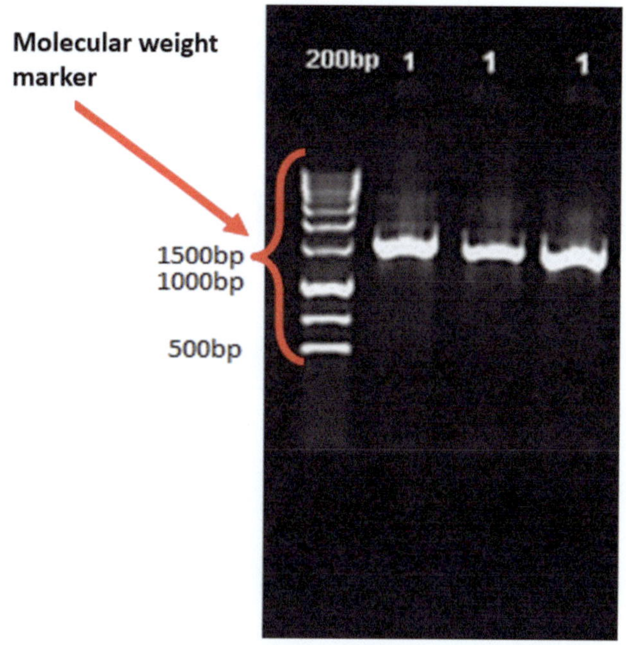

Fig. 7 Amplification of a specific 1500 bp band generated using primers FD1 and RP2 from FSO1 lactic acid bacteria

3.2 Molecular Validation of 16S rDNA PCR Amplification

Amplification of the 16S rDNA gene was carried out on three DNA samples extracted from *L. plantarum* FSO1 strain. The products of the PCR were purified and confirmed by agarose gel electrophoresis with a 200 bp molecular weight marker. The results are presented in Fig. 7. The electrophoretic run of the amplified product was 20 min. A distinct band at 1500 bp was observed for the DNA of the strain analyzed (Fig. 7) which indicated amplification. As shown in the results, all of the 16S.

The purified products of amplification of the 16S rDNA gene of the FSO1 lactic acid bacterium were sequenced by the UATRS-CNRST Rabat molecular biology laboratory.

The sequences of the 16S rDNA gene of FSO1 lactic acid bacterium generated by the ABI PRISM 3130 xl Genetic Analyzer were compared with the help of Sequencing Analysis v5.3.1 software.

Homology search of the processed sequences was done by BLAST (NCBI tool). BLAST analysis (https://blast.ncbi.nlm.nih.gov/Blast.cgi) is a computer software tool used for DNA sequence alignment and comparison with those in the GenBank database. The 16S rDNA sequence analysis results of the *L. plantarum* FSO1 strain validated its preliminary identification by the API50 CHL system (Table 3).

Table 3 Biochemical and molecular identification of the *L. plantarum* FSO1 strain

Strains	Biochemical identification (API 50CHL)	Molecular identification (16S rDNA)			
		Identification	Length of query	% of identification (%)	Number of accession
FSO1	*L. plantarum*	*L. plantarum*	1502	100	NR_115605.1

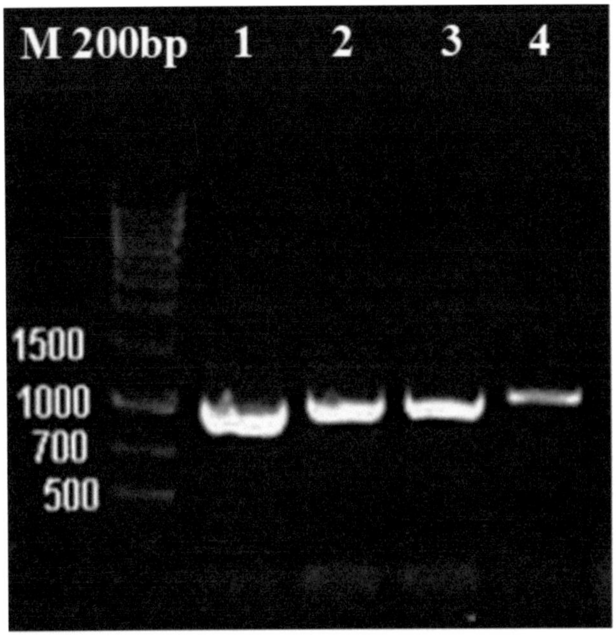

Fig. 8 PCR amplification results demonstrating the presence of the esterase gene. *Legend* M, molecular weight marker; (1) *L. plantarum* FSO1; (2) *C. pelliculosa* L18; (3 and 4) oleuropeinolytic strain (Ghabbour et al. 2011)

3.3 Detection of Esterase Genes

PCR amplification using specially designed primers for the esterase genes produced a single 1000 bp band for the esterase gene (Fig. 8). The presence of the esterase gene was detected in all of the strains tested.

Gene sequence analysis revealed a high degree of homology to *L. plantarum* sequences deposited in GenBank, and nucleotide sequence identity was above 100%.

By using molecular techniques, we verified the existence of esterase genes in some L. species that participate in table olive fermentation. Here, we screened some L. species for the existence of these esterase genes.

The results of the present work prove that table olive lactobacilli, besides having the capacity to be lactic fermentation inducers, are also an excellent source of enzymes that may help in positively modulating the flavor of the olives. Although a number of strains have genes capable of acting on some substrates, it is important to follow through and examine whether the genes are expressed during conditions of lactic fermentation and also ascertain whether the resultant enzymes are active in natural olive brines. The research therefore provides a foundation for identifying possible strains for further research regarding gene expression and enzyme activity.

4 Conclusion

The current study confirms that *L. plantarum* FSO1 and *C. pelliculosa* L18 produce extracellular esterases that are most active at pH 6 and 0% NaCl. The enzymes hold great potential for debittering green olives, offering a treatment process that is chemical free. Large-scale production of the enzyme and its use in industrial olive fermentation must be investigated in the future studies. Microbial enzymes are more multi-functional, inexpensive, and easier to obtain than those from animal or plant sources. The sensitivity to different physical and chemical characteristics was utilized to establish the identification of esterases. The esterase production and tolerance to stress conditions were also shown by lactic acid bacteria and yeast strains. Stress resistance is a determinant that *L. plantarum* FSO1 and *C. pelliculosa* L18 are appropriate to be applied for biotechnological purposes.

Our knowledge of the production and active condition of the most important enzyme in the process of industrial table olive fermentation has been further explored in this work.

References

Ahmadova, A., S.D. Todorov, Y. Choiset, H. Rabesona, T. Mirhadi Zadi, A. Kuliyev, B.D.G.d.M. Franco, J.-M. Chobert, and T. Haertlé, *Evaluation of antimicrobial activity, probiotic properties and safety of wild strain Enterococcus faecium AQ71 isolated from Azerbaijani Motal cheese.* Food Control, 2013. **30**(2): p. 631–641. https://doi.org/10.1016/j.foodcont.2012.08.009

Alfano, G., G. Lustrato, G. Lima, D. Vitullo, and G. Ranalli, *Characterization of composted olive mill wastes to predict potential plant disease suppressiveness.* Biological Control, 2011. **58**(3): p. 199–207. https://doi.org/10.1016/j.biocontrol.2011.05.001

Anagnostopoulos, D.A., E. Kamilari, and D. Tsaltas, *Evolution of bacterial communities, physicochemical changes and sensorial attributes of natural whole and cracked picual table olives during spontaneous and inoculated fermentation.* Frontiers in microbiology, 2020. **11**: p. 1128. https://doi.org/10.3389/fmicb.2020.01128

Bonatsou, S., A. Benitez, F. Rodriguez-Gomez, E.Z. Panagou, and F.N. Arroyo-Lopez, *Selection of yeasts with multifunctional features for application as starters in natural black table olive processing.* Food Microbiol, 2015. **46**: p. 66–73. https://doi.org/10.1016/j.fm.2014.07.011

Charoenprasert, S. and A. Mitchell, *Factors influencing phenolic compounds in table olives (Olea europaea).* Journal of Agricultural and Food Chemistry, 2012. **60**(29): p. 7081–7095. https://doi.org/10.1021/jf3017699

Clarridge, J.E., *Impact of 16S rRNA gene sequence analysis for identification of bacteria on clinical microbiology and infectious diseases.* Clinical microbiology reviews, 2004. **17**(4): p. 840–862. https://doi.org/10.1128/CMR.17.4.840-862.2004

Fayek, N.M., M.A. Farag, and F.R. Saber, *Metabolome classification via GC/MS and UHPLC/MS of olive fruit varieties grown in Egypt reveal pickling process impact on their composition.* Food Chemistry, 2021. **339**: p. 127861. https://doi.org/10.1016/j.foodchem.2020.127861

Ghabbour, N., Z. Lamzira, P. Thonart, P. Cidalia, M. Markaoui, and A. Asehraou, *Selection of oleuropein-degrading lactic acid bacteria strains isolated from fermenting Moroccan green olives.* Grasas y Aceites, 2011. **62**(1): p. 84–89. https://doi.org/10.3989/gya.055510

Gikas, E., N. Papadopoulos, and A. Tsarbopoulos, *Kinetic study of the acidic hydrolysis of oleuropein, the major bioactive metabolite of olive oil.* Journal of liquid chromatography & related technologies, 2006. **29**(4): p. 497–508. https://doi.org/10.1080/10826070500474113

Harris, K.A., K.J. Fidler, J.C. Hartley, J. Vogt, N.J. Klein, F. Monsell, and V.M. Novelli, *Unique case of Helicobacter sp. osteomyelitis in an immunocompetent child diagnosed by broad-range 16S PCR.* Journal of clinical microbiology, 2002. **40**(8): p. 3100–3103. https://doi.org/10.1128/JCM.40.8.3100-3103.2002

Laemmli, U.K., *Cleavage of Structural Proteins during the Assembly of the Head of Bacteriophage T4.* Nature, 1970. **227**(5259): p. 680–685. https://doi.org/10.1038/227680a0

Mackness, M.I., C.H. Walker, D.G. Rowlands, and N.R. Price, *Esterase activity in homogenates of three strains of the rust red flour beetle Tribolium castaneum (Herbst).* Comparative Biochemistry and Physiology Part C: Comparative Pharmacology, 1983. **74**(1): p. 65–68. https://doi.org/10.1016/0742-8413(83)90150-0

Mtshali, P.S., B. Divol, P. Van Rensburg, and M. Du Toit, *Genetic screening of wine-related enzymes in Lactobacillus species isolated from South African wines.* Journal of Applied Microbiology, 2010. **108**(4): p. 1389–1397. https://doi.org/10.1111/j.1365-2672.2009.04535.x

Principi, P., G. Ranalli, F. da Borso, M. Pin, E. Zanardini, and C. Sorlini, *Microbiological aspects of humid husk composting.* Journal of Environmental Science and Health, Part B, 2003. **38**(5): p. 645–661. https://doi.org/10.1081/PFC-120023521

Rokni, Y., *Contribution à l'étude de la biodégradation de l'oleuropéine par des souches de bactéries lactiques et de levures isolées des olives vertes en fermentation naturelle.* 2017, Université Mohammed Premier, Faculté des Sciences - Oujda.

Rokni, Y., N. Ghabbour, N.-E. Chihib, P. Thonart, and A. Asehraou, *Physico-chemical and microbiological characterization of the natural fermentation of Moroccan Picholine green olives variety.* Journal of Materials and Environmental Science, 2015. **6**(6): p. 1740–1751.

Rokni, Y., H. Abouloifa, R. Bellaouchi, S. Gaamouche, K. Mchiouer, I. Hasnaoui, Z. Lamzira, N. Ghabbour, and A. Asehraou, *Technological process of fermented olive.* Arabian Journal of Chemical and Environmental Research, 2017. **04**(2): p. 63–91.

Rokni, Y., H. Abouloifa, R. Bellaouchi, I. Hasnaoui, S. Gaamouche, Z. Lamzira, R. Salah, E. Saalaoui, N. Ghabbour, and A. Asehraou, *Characterization of beta-glucosidase of Lactobacillus plantarum FSO1 and Candida pelliculosa L18 isolated from traditional fermented green olive.* J Genet Eng Biotechnol, 2021. **19**(1): p. 117. https://doi.org/10.1186/s43141-021-00213-3

Soliman, N.A., M. Knoll, Y.R. Abdel-Fattah, R.D. Schmid, and S. Lange, *Molecular cloning and characterization of thermostable esterase and lipase from Geobacillus thermoleovorans YN isolated from desert soil in Egypt.* Process Biochemistry, 2007. **42**(7): p. 1090–1100. https://doi.org/10.1016/j.procbio.2007.05.005

Vaccalluzzo, A., L. Solieri, D. Tagliazucchi, A. Cattivelli, S. Martini, A. Pino, C. Caggia, and L. Randazzo Cinzia, *Metabolomic and Transcriptional Profiling of Oleuropein Bioconversion into Hydroxytyrosol during Table Olive Fermentation by Lactiplantibacillus plantarum.* Applied and Environmental Microbiology, 2022. **88**(6): p. e02019–21. https://doi.org/10.1128/aem.02019-21

Weisburg, W.G., S.M. Barns, D.A. Pelletier, and D.J. Lane, *16S ribosomal DNA amplification for phylogenetic study.* Journal of bacteriology, 1991. **173**(2): p. 697–703. https://doi.org/10.1128/jb.173.2.697-703.1991

Wray, W., T. Boulikas, V.P. Wray, and R. Hancock, *Silver staining of proteins in polyacrylamide gels.* Analytical Biochemistry, 1981. **118**(1): p. 197–203. https://doi.org/10.1016/0003-2697(81)90179-2

MIX
Papier aus verantwortungsvollen Quellen
Paper from responsible sources
FSC® C105338

If you have any concerns about our products,
you can contact us on
ProductSafety@springernature.com

In case Publisher is established outside the EU,
the EU authorized representative is:
Springer Nature Customer Service Center GmbH
Europaplatz 3, 69115 Heidelberg, Germany

Printed by Libri Plureos GmbH
in Hamburg, Germany